PROGRAMMING AND CUSTOMIZING THE PICAXE MICROCONTROLLER

MW00396397

PROGRAMMING AND CUSTOMIZING THE PICAXE MICROCONTROLLER

SECOND EDITION

DAVID LINCOLN

New York   Chicago   San Francisco   Lisbon   London   Madrid
Mexico City   Milan   New Delhi   San Juan   Seoul
Singapore   Sydney   Toronto

The McGraw-Hill Companies

Cataloging-in-Publication Data is on file with the Library of Congress

Thank you to all who worked on this book to bring it to its printed form.

Copyright © 2011, 2006 by The McGraw-Hill Companies, Inc. All rights reserved. Printed in the United States of America. Except as permitted under the United States Copyright Act of 1976, no part of this publication may be reproduced or distributed in any form or by any means, or stored in a data base or retrieval system, without the prior written permission of the publisher.

1 2 3 4 5 6 7 8 9 0   DOC/DOC   1 6 5 4 3 2 1 0

ISBN 978-0-07-174554-3
MHID 0-07-174554-8

Sponsoring Editor: Judy Bass
Editing Supervisor: Stephen M. Smith
Production Supervisor: Pamela A. Pelton
Acquisitions Coordinator: Michael Mulcahy
Project Manager: Shikha Sharma, Aptara, Inc.
Copy Editor: Sheila Johnston
Proofreader: Shashi Lal Das, Aptara, Inc.
Indexer: Amit Srivastava, Aptara, Inc.
Art Director, Cover: Jeff Weeks
Composition: Aptara, Inc.

Printed and bound by RR Donnelley.

McGraw-Hill books are available at special quantity discounts to use as premiums and sales promotions, or for use in corporate training programs. To contact a representative, please e-mail us at bulksales@mcgraw-hill.com.

This book is printed on acid-free paper.

Information contained in this work has been obtained by The McGraw-Hill Companies, Inc. ("McGraw-Hill") from sources believed to be reliable. However, neither McGraw-Hill nor its authors guarantee the accuracy or completeness of any information published herein, and neither McGraw-Hill nor its authors shall be responsible for any errors, omissions, or damages arising out of use of this information. This work is published with the understanding that McGraw-Hill and its authors are supplying information but are not attempting to render engineering or other professional services. If such services are required, the assistance of an appropriate professional should be sought.

About the Author

David Lincoln has spent more than 40 years in the IT and telecommunication industries, and many corporations have run their businesses using custom software he has designed and written. He has now moved out of the corporate area and runs Microzed Computers, the Australian distributor for PICAXE products. David also provides consulting and professional development services to manufacturers, schools, teachers, and other interested parties.

Disclaimer

The material presented in this book is provided for the purpose of performing experiments in the use of digital electronics and microcontroller programming. The circuits and code presented are for experimental purposes only, and the author does not warrant them to be suitable for any particular purpose.

Trademarks

IBM is a registered trademark of International Business Machines Corporation.

PICAXE is a registered trademark licensed to Revolution Education Ltd. by Microchip Technology Inc. (Revolution Education Ltd. is not an agent or representative of Microchip Technology Inc. and has no authority to bind Microchip Technology Inc. in any way. The PICAXE product is developed and distributed by Revolution Education Ltd.)

Programming Editor is a registered trademark of Revolution Education Ltd.

PIC and PICmicro are registered trademarks of Microchip Technology Inc. in the United States and other countries.

Microsoft Windows is a registered trademark of Microsoft Corporation.

Macintosh is a registered trademark of Apple Computer Inc.

iPhone is a registered trademark of Apple Corporation.

1-Wire is a registered trademark of Dallas Semiconductor Corporation.

I^2C Bus is a registered trademark of Philips Corporation.

SPI Bus, named by Motorola, Inc.

UNI/O is a registered trademark of Microchip Technology Inc.

Wikipedia is a registered trademark of the not-for-profit Wikimedia Foundation.

References

PICAXE Manual Sections 1, 2, and 3, copyright Revolution Education Ltd.

Various datasheets and/or application notes pertaining to the electronic components used in the experiments, which are copyrighted by their owners.

Wikipedia, the free encyclopedia.

CONTENTS

PROGRAMMING AND CUSTOMIZING THE PICAXE MICROCONTROLLER

INTRODUCTION

What Is a Microcontroller?

Microcontrollers (often abbreviated MCU or µC) are similar to the processor in a personal computer (PC), except on a much smaller scale. Since they take many different forms, it is difficult to provide one definition that covers them all.

In general, a microcontroller is an integrated circuit (chip) that contains a processor, memory, input ports, and output ports. The term "microcontroller" tends to imply that all these components are included in a single chip.

Microcontrollers may be packaged in dual-in-line package (DIL), surface-mount packages (SOIC), or any of the integrated circuit packages that are available. Figure 1.1 shows the general arrangement of a microcontroller.

Microcontrollers are now available in a wide range of sizes and architectures. The smallest costs less than $1 each in production quantities, whereas the largest are comparable in power to the processor used in the first PCs of the 1980s.

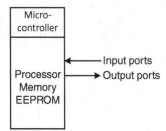

Figure 1.1 Microcontroller architecture.

Where Are Microcontrollers Used?

Microcontrollers are used in most electronic devices. The list includes toys, home appliances, television sets, microwave ovens, washing machines, video players, mobile phones, cars, aeroplanes, smart battery chargers, security systems, digital cameras, robots, smart cards, disk drives, network routers, MP3 players, and many more.

The Advantages of a Programmable Device

Microcontrollers are programmable devices, which means that there is a set of instructions within them, called a program or firmware that controls their behavior. The instructions are specific to a particular application and the same microcontroller chip can be configured for different applications by programming it with a different set of instructions.

Modern microcontroller-based devices are often designed such that the microcontroller can be programmed after the circuit is manufactured. This allows the user to install the latest firmware–thus keeping the device up to date.

Why Is It Important to Understand Microcontrollers?

In today's world, more and more devices are becoming microcontroller based and the people who design, maintain, configure, and use these devices will be required to have an understanding of microcontrollers.

What Is a PICAXE® Microcontroller?

A PICAXE microcontroller (Fig. 1.2) is a bootstrap-coded microcontroller that runs on a selection of Microchip's line of PIC® microcontrollers. The low cost of PICAXE

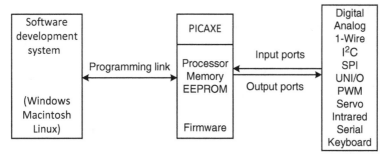

Figure 1.2 PICAXE system architecture.

chips, combined with their simplicity and free development tools, makes them ideal for learning tools, proof of concept, and many other applications. Students, hobbyists, technicians, farmers, schools, government departments, and manufacturers commonly use them.

HANDLING PRECAUTIONS

PICAXE chips are manufactured from metal-oxide-semiconductor (MOS) technology, which is subject to damage by static electricity. Modern MOS devices have built-in electrostatic-discharge (ESD) protection; however, it is wise to use antistatic procedures when handling MOS devices, especially in low-humidity environments.

The Experiments

The early experiments in this book are simple in nature and are designed to be implemented with the Schools Experimenter board, which is manufactured by Revolution Education and available from PICAXE distributors worldwide as part AXE092. The Schools Experimenter board carries an 8-pin PICAXE chip, three light-emitting diodes (LED's), a press button switch, and a light-dependent resistor (LDR). There is also a connector to allow external components to be easily connected and switches that allow on-board circuitry to be isolated from the chip. It is, therefore, an ideal platform for performing the initial experiments.

For the benefit or readers who prefer to prototype their own circuits, the circuit for the Schools Experimenter board is given later in this book. More advanced experiments make use of 18-, 20-, 28-, and 40-pin chips. Revolution Education make a range of circuit boards to suit these chips and for readers who prefer to construct their own prototypes, the full circuit of each experiment is given.

The early experiments need little knowledge of programming or electronics and are accompanied with a full description of each line of code. It is assumed that readers will have some knowledge of programming and electronics by the time they attempt the more advanced experiments and the code descriptions are only given where the code has not already been explained or is not self-explanatory.

With the circuits shown in this book, it is almost impossible to cause damage to the PICAXE chip and readers are encouraged to experiment for themselves. Some care must be taken if making changes to the circuits themselves. In particular, voltage ranges should be observed and output pins should not be connected together or short circuited. Resistance values should not be reduced to the point where they are too low to limit currents to safe values; if in doubt, leave resistance values as they are.

PICAXE ARCHITECTURE

Chip Architecture

At the time of writing, PICAXE chips are available in "M," "X," "X1," and "X2" series. Revolution Education is constantly in the process of developing new and improved chips and there may be new chips added to the range by the time you read this book.

TABLE 2.1   SUMMARY OF PICAXE CHARACTERISTICS

CHIP	PROGRAM MEMORY (APPROX. LINES OF CODE)	VARI-ABLES (BYTES)	ADDITIONAL MEMORY (SCRATCHPAD/ STORAGE BYTES)	EEPROM (BYTES)	I/O PINS	ANALOG PINS	SUPPLY VOLTAGE (V)
08M	80–220	14	0/48	256-program	5	3	4.2–5.5
14M	80–220	14	0/48	256-program	11	2 + 3	4.2–5.5
18M	80–220	14	0/48	256-program	13	3	4.2–5.5
20M	80–220	14	0/48	256-program	16	4	4.2–5.5
18X	600–1,800	14	0/96	256	13	3	4.2–5.5
28X1	2,000–3,200	28	128/95	256	20	4	4.2–5.5
40X1	2,000–3,200	28	128/95	256	31	7	4.2–5.5
20X2	2,000–3,200	56	128/72	256	17	11	1.8–5.5
28X2	2,000–3,200	56	1,024/200	256	21	9	4.2–5.5
28X2–3 V	2,000–3,200	56	1,024/200	256	21	9	1.8–3.6
40X2	2,000–3,200	56	1,024/200	256	32	12	4.2–5.5
40X2–3 V	2,000–3,200	56	1,024/200	256	32	12	1.8–3.6

The "M-" series chips are the entry-level chips and are available in 8-, 14-, 18-, and 20-pin packages. The "X" series is available in an 18-pin package. The "X1" series are available in 28- and 40-pin packages, and the "X2" series are available in 20-, 28-, and 40-pin packages. Table 2.1 summarizes some characteristics of each chip.

Powering the PICAXE

The voltage requirements for powering PICAXE chips fall into two categories; standard and low voltage. The exception is the 20X2, which will operate from both standard- and low-voltage supplies. PICAXE chips may be supplied from any form of power supply, such as batteries or a mains-powered rectifier, filter, and regulator

The recommended supply voltage for the standard-voltage chips is in the range 4.2 to 5.5 V and this may be supplied from batteries or a mains-powered rectifier, filter, and regulator. When the supply is derived from a voltage regulator, 5 V is commonly used.

The low-voltage chips will run from supply voltages in the range 1.8 to 3.3 V. Low-voltage chips should only be programmed using a USB programming cable or USB breakout programmer. The voltage from an RS232 port (up to + or −12 V) can permanently damage low-voltage chips. Note that the voltage applied to the input pins of any PICAXE chip must not exceed the supply voltage by more than 0.3 V.

The PICAXE-20X2 can be powered from supply voltages in the range 1.8 to 5.5 V. The voltage on the input pins of the 20X2 must not exceed the supply voltage by more than 0.3 V; inputs can accept 5 V if the supply voltage is 5 V.

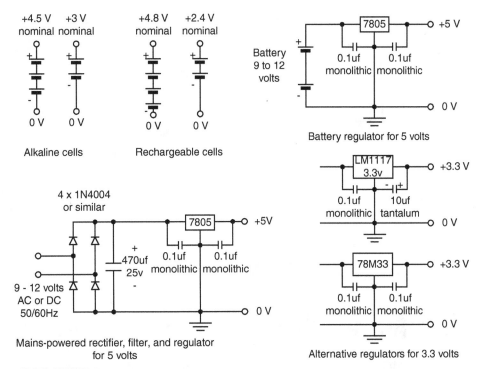

Figure 2.1 Powering PICAXE chips.

When running standard PICAXE chips from batteries, the recommended configurations are 3 × AA alkaline cells or 4 × AA rechargeable cells. Low-voltage chips can be run from 2 × AA cells, either alkaline or rechargeable. Be warned, however, that freshly charged rechargeable cells can present higher than nominal voltages (up to 1.6 V per cell) for short periods of time, especially if they have been charged with a fast charger; this may exceed the recommended voltage range for the chip. When running from regulated power supplies, 5 V is recommended for standard- and 3.3 V for low-voltage chips, although any voltage within the operating range can be used. Suitable power supplies are shown in Fig. 2.1.

Resetting the PICAXE

All PICAXE chips can be reset by removing and restoring power; this is the only method of resetting chips that do not have an external reset pin. Chips that have an external reset pin can be reset by bringing the reset pin to a logic low level. The reset pin is active low and for normal operation must be held at the logic high level. The circuit in Fig. 2.2 is suitable for resetting PICAXE chips that have a reset pin. X1- and X2-series chips may also be reset by using the reset command.

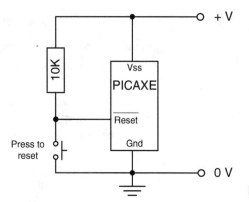

Figure 2.2 PICAXE reset circuit.

Resetting a PICAXE chip will restart the program and initialize all variables and ports to the start-up values; the program is not erased and the contents of EEPROM are not affected.

Downloading Programs to the PICAXE

The PICAXE programming circuit consists of two resistors, as shown in Fig. 2.3b. The program download system is a simple matter of connecting a cable between the development computer and the PICAXE. There is also the facility to extend the programming link over a TCP/IP network or use the USB breakout programmer that can operate as a programming cable or stand-alone programmer. Figure 2.3a shows the methods of programming PICAXE.

The USB cable (part AXE027) is currently the preferred cable for downloading programs, because it works with all supported computers and all PICAXE chips. The

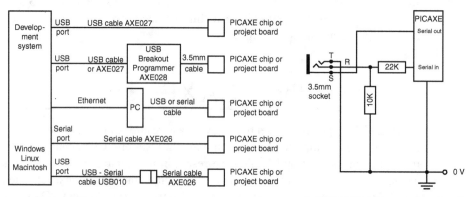

a. PICAXE program download.   b. PICAXE programming circuit.

Figure 2.3 PICAXE program download (a) and programming circuit (b).

serial cable (part AXE026) is a lower cost option to the USB cable; however, many modern computers do not have a serial port and it is not suitable for programming low-voltage PICAXE chips, because the RS232 voltages can damage them.

The USB breakout programmer (part AXE028) can function as a USB programming cable or can be used as a stand-alone programmer. The stand-alone programmer facility can be a big advantage when there is a number of PICAXEs to be programmed with the same program.

Programs can also be downloaded over TCP/IP networks. Two computers are needed on the same network, one acting as a master and the other acting as a slave. Instructions for TCP/IP configuration can be found by selecting the "PICAXE | Wizards | COM to TCP/IP" options from the programming editor menu.

Clocking PICAXE Chips

All current PICAXE chips have an internal clock that defaults to 4 MHz for the M-, X-, and X1-series, and 8 MHz for X2-series chips. An external 3-pin resonator can optionally be connected to the X1- and X2-series chips; the X2-series chips also support 2-pin resonators and crystals. The X2-series chips make use of an internal four times multiplier for the clock and the actual clock frequency is four times the resonator frequency. Table 2.2 shows the default and standard frequencies.

In most applications, the clock frequency can be chosen from the standard frequencies. In some specialized applications, it may be necessary to use an external resonator that is not one of the standard frequencies as this may cause issues when programming the chip. If this is the case, then the resonator can be removed during programming. X1 and X2 chips will revert to their default frequency for program downloads when

TABLE 2.2 CLOCK AND RESONATOR FREQUENCIES

CHIP	DEFAULT FREQUENCY (MHz)	INTERNAL CLOCK FRE-QUENCIES (MHz)	EXTERNAL RESONATOR FREQUENCIES (MHz)	CLOCK FREQUENCIES WITH EXTERNAL RESONATOR (MHz)
08M	4	4, 8		
14M	4	4, 8		
18M	4	4, 8		
20M	4	4, 8		
18X	4	4, 8		
28X1	4	4, 8	4, 8, 16, other	4, 8, 16, other
40X1	4	4, 8	4, 8, 16, other	4, 8, 16, other
20X2	8	4, 8, 16, 32, 64		
28X2	8	4, 8	4, 8, 10, other	16, 32, 40, other × 4
40X2	8	4, 8	4, 8, 10, other	16, 32, 40, other × 4

running from their internal resonator. Other chips will use the last clock frequency that was programmed into the chip by a **setfreq** command.

There are also some commands that make use of the clock for timing purposes and those commands may be unusable or may operate differently if one of the standard frequencies in Table 2.2 is not used, e.g., **serin** and **pause**. There is a complete list of commands affected by resonator frequency in the PICAXE manual, Section 1.

The clock frequency for any PICAXE chip that is operating from its internal resonator can be changed to one of the frequencies shown in Table 2.2 by issuing a **setfreq** command. When an external resonator is used, the clock frequency is determined by the external resonator.

At a clock speed of 4 MHz, a PICAXE will execute approximately 1,000 lines of code per second. This rate increases proportionally with an increase in clock speed.

PICAXE Memory

PICAXE memory is divided into the following categories:

- General purpose
- System
- Scratchpad
- Storage
- Special function
- EEPROM
- TABLE

The general-purpose, system, scratchpad, storage, and special-function variables are volatile and will lose their contents when power is removed from the chip or a reset occurs. EEPROM and TABLE memory are nonvolatile and will retain their contents until overwritten either by program command or when a chip is reprogrammed.

The general-purpose, scratchpad, and storage variables are available as data storage. The amount of storage varies depending on the PICAXE chip. The system and special function variables are dedicated to system functions, such as, flags, input/output ports, and pointers, although some can also be used for data storage if the particular system function is not being used. Table 2.3 summarizes the data storage.

GENERAL-PURPOSE VARIABLES

General-purpose variables are variables that can be used in commands wherever a variable is required. They can be addressed as words, bytes, or bits. However, it should be noted that the word variables are made up of two-byte variables and the bit variables are the individual bits of the first two-byte variables (first four for X1 and X2 series) and the first word variable (first two for X1 and X2 series).

The primary command for operating on variables is the **let** command, although there are many other commands that read from, or place values in, variables.

TABLE 2.3 SUMMARY OF PICAXE DATA STORAGE

	GENERAL-PURPOSE VARIABLES			SCRATCHPAD (BYTES)	STORAGE VARIABLES (BYTES)	EEPROM (BYTES)
CHIP	WORDS	BYTES	BITS			
08M	7	14	16		48	256 minus program size
14M	7	14	16		48	256 minus program size
18M	7	14	16		48	256 minus program size
20M	7	14	16		48	256 minus program size
18X	7	14	16		96	256
28X1	14	28	32	128	95	256
40X1	14	28	32	128	95	256
20X2	28	56	32	128	72	256
28X2	28	56	32	1,024	200	256
40X2	28	56	32	1,024	200	256

Bit variables consist of a single binary digit that can hold the value 1 (one) or the value 0 (zero). A bit cannot exist by itself, but is always part of a byte or a word. Bits can be set, reset, and isolated by using the **let** command in conjunction with Boolean functions; **setbit** and **clearbit** commands are available for X1- and X2-series chips.

Byte variables consist of eight bits (binary digits) and can hold an ASCII character, an unsigned binary number in the range 0 to 255, a signed binary number in the range -128 to $+127$, two BCD digits, or any other data that requires up to eight bits for storage, e.g., the state of an input port.

Word variables consist of 16 bits (binary digits) and can hold two ASCII characters, an unsigned binary number in the range 0 to 65,535, a signed binary number in the range $-32,768$ to $+32,767$, four BCD digits, or any other data that requires up to 16 bits for storage, e.g., 10-bit analog reading.

Table 2.4 shows the relationship between word and byte variables; Table 2.5 shows the relationship between word, byte, and bit variables.

Any changes a program makes to a bit, byte, or word variable will affect the other variables that occupy the same space. For example, it can be seen from Table 2.4 that word variable w2 occupies the same space as byte variables b5 and b4 and any changes a program makes to byte variables b5 or b4 will affect word variable w2 (and vice versa). From Table 2.5 it can be seen that bit variable bit7 is part of byte variable b0 and word variable w0, and bit variable bit8 is part of byte variable b1 and word variable w0. Any changes made to bit7 or bit8 will affect b0, b1, and w0 (and vice versa). Note also that variable names are not case sensitive and W0 and w0 refer to word 0 (zero), B1 and b1 refer to byte 1, and BIT2, Bit2, and bit2 refer to bit 2.

TABLE 2.4   RELATIONSHIP BETWEEN WORD AND BYTE VARIABLES

WORD VARIABLE	MOST SIGNIFICANT BYTE	LEAST SIGNIFICANT BYTE
W0	B1	B0
W1	B3	B2
W2	B5	B4
W3	B7	B6
W4	B9	B8
W5	B11	B10
W6	B13	B12
Similarly for X1 series W7–W13	B15–B27	B14–B26
Similarly for X2 series W14–W27	B29–B55	B28–B54

SYSTEM VARIABLES

System variables are reserved for use by the PICAXE firmware and the firmware may place values into system variables during program execution. There are eight word variables and one-byte variable available in the X1 and X2 parts. If a system variable is not being used by the system, then a program may use that system variable for storage.

TABLE 2.5   RELATIONSHIP BETWEEN WORD, BYTE, AND BIT VARIABLES

WORD VARIABLE	BYTE VARIABLE	BIT VARIABLE
W1, bit 0 (least significant)	B0, bit 0 (least significant)	Bit0
W1, bit 1	B0, bit 1	Bit1
W1, bit 2	B0, bit 2	Bit2
W1, bit 3	B0, bit 3	Bit3
W1, bit 4	B0, bit 4	Bit4
W1, bit 5	B0, bit 5	Bit5
W1, bit 6	B0, bit 6	Bit6
W1, bit 7	B0, bit 7 (most significant)	Bit7
W1, bit 8	B1, bit 0 (least significant)	Bit8
W1, bit 9	B1, bit 1	Bit9
W1, bit 10	B1, bit 2	Bit10
W1, bit 11	B1, bit 3	Bit11
W1, bit 12	B1, bit 4	Bit12
W1, bit 13	B1, bit 5	Bit13
W1, bit 14	B1, bit 6	Bit14
W1, bit 15 (most significant)	B1, bit 7 (most significant)	Bit15
Similarly for X1 and X2 series, W2	B2–B3	Bit16–Bit31

The system variables are documented in the PICAXE manual and are reproduced here for convenience.

s_w0		Reserved for future use
s_w1		Reserved for future use
s_w2		Reserved for future use
s_w3	timer3	timer3 value (X2 only)
s_w4	compvalue	comparator results (X2 only)
s_w5	hserptr	hardware serin pointer
s_w6	hi2clast	hardware hi2c last byte written (slave mode)
s_w7	timer	timer value

flags byte

bit 0	flag0	hint0flag	X2 only—interrupt on B.0
bit 1	flag1	hint1flag	X2 only—interrupt on B.1
bit 2	flag2	hint2flag	X2 only—interrupt on B.2
bit 3	flag3	hintflag	X2 only—interrupt on B.0, B.1, or B.2 has occurred
bit 4	flag4	compflag	X2 only—occurs on comparator change
bit 5	flag5	hserflag	hserial background receive has occurred
bit 6	flag6	hi2cflag	hi2c write has occurred (slave mode)
bit 7	flag7	toflag	timer overflow flag

SCRATCHPAD

The scratchpad is an area of memory that can be used for the storage of data and arrays. It is accessed by means of the **put** command, **get** command, and the pointer **ptr**. The PICAXE-28X1, -40X1, and -20X2 have 128 bytes of scratchpad; the PICAXE-28X2 and -40X2 have 1,024 bytes of scratchpad. The **ptr** method of accessing the scratchpad is especially useful, since it allows a scratchpad variable to be used in most places that a general-purpose variable can be used. Pointers are discussed in detail elsewhere in this book.

STORAGE VARIABLES

Storage variables are byte variables that can be used to hold data. They are accessed by the **peek** and **poke** commands and on X2 parts can be accessed via the pointer **bptr**. Table 2.6 summarizes the storage variables.

For the X2 parts, the byte variables (b0–b55) and storage variables form a contiguous block of storage that can be addressed by the **peek** and **poke** commands and the byte scratchpad pointer **bptr**. Figure 2.5 (see later) shows the relationship.

SPECIAL FUNCTION

Special-function variables are variables that perform extra functions, such as the state of an input/output port, the direction of an input/output pin, or a pointer to another memory location. The special-function variables are summarized in Table 2.7.

TABLE 2.6 PICAXE STORAGE VARIABLES		
CHIP	BYTES	ADDRESSES
08M	48	80 to 127
14M	48	80 to 127
18M	48	80 to 127
20M	48	80 to 127
18X	48	80 to 127
	48	192 to 239
28X1	47	80 to 126
	48	192 to 239
40X1	47	80 to 126
	48	192 to 239
20X2	72	56 to 127
28X2	200	56 to 255
40X2	200	56 to 255

EEPROM

All PICAXE chips have Electrically Erasable Programmable Read-Only Memory that can be used for storing data or constants, The amount and type of EEPROM available varies depending on the chip and, in some cases, with the amount of memory used by program code. The big advantage of EEPROM is that the contents are retained when power is removed or the chip is reset.

The "M-" series chips store their program code in EEPROM and this reduces the amount of EEPROM space available for a program to use. The amount of EEPROM space available for an "M-" series chip can be calculated by downloading or syntax-checking a program and subtracting the number of bytes of memory used from 256. The first EEPROM address that can be used is next address after the end of the program. For example, a PICAXE-08M program is syntax-checked and gives the following message "Memory used = 28 bytes out to 256" This means that the PICAXE-08M has 256 bytes of EEPROM and 28 of those bytes (0 to 27) have been used by the program code. Therefore, 228 bytes of EEPROM (256 – 28) are available beginning at address 28.

The commands that operate on EEPROM are:

data

eeprom

read

write

TABLE 2.7   SPECIAL-FUNCTION VARIABLES

SPECIAL-FUNCTION VARIABLE	08M	14M	18M	20M	18X	X1 SERIES	X2 SERIES
Pins	✓	✓	✓	✓	✓	✓	
Dirs	✓						
Outpins	✓	✓	✓	✓	✓	✓	
Infra	✓	✓	✓	✓	✓		
keyvalue			✓		✓		
Flags						✓	✓
pinsA							✓
dirsA							✓
pinsB							✓
dirsB							✓
pinsC		✓				✓	✓
dirsC		✓				✓	✓
pinsD							✓
dirsD							✓
ptr						✓	✓
ptrl						✓	✓
ptrh						✓	✓
@ptr						✓	✓
@ptrinc						✓	✓
@ptrdec						✓	✓
Bptr							✓
@bptr							✓
@bptrinc							✓
@bptrdec							✓

The syntax of the **data** and **eeprom** commands is:

data {location,} (data{,data} ...)

eeprom {location,} (data{,data} ...)

where

location is the EEPROM address that receives the first data byte.

data is the data to be stored.

location is optional and if not present, the data will be stored in the next location, or, if no previous data has been specified, the location defaults to 0 (zero).

The **data** and **eeprom** commands are functionally identical and are used to load data into the EEPROM at the time the PICAXE is being programmed. These commands do not take any space in the PICAXE program memory and on the X-, X1-, and X2-series chips, the **#no_data** directive can be used to prevent the EEPROM from being loaded.

Examples:

eeprom (1)	Load the value 1 (one) into the next EEPROM location when the PICAXE is programmed. The next EEPROM address is used because the location is not specified in the command and will default to 0 (zero) if no previous **eeprom** or **data** commands have been issued.
data 10, (7, 6, 5)	Load the value 7 into EEPROM location 10, the value 6 into location 11, and the value 5 into location 12 when PICAXE is programmed.
data 20, ("ABC", 8)	Load the ASCII characters "A," "B," and "C" into EEPROM locations 20, 21, and 22, and the value 8 into location 23 when PICAXE is programmed.

The syntax of the **read** and **write** commands is:

read location, {WORD} variable {, variable . . . }

write location, {WORD} variable {, variable . . . }

where

location is the EEPROM address that receives the first data byte *data* is the data to be stored.

WORD is an optional keyword and, if present, specifies that the following variables are words rather than bytes. Words are stored with low byte of the word at the lower address and the high byte of the word at the higher address.

Examples:

read 10, b0	This command will read a byte from EEPROM location 10 and place the result into variable b0.
read 15, b2, b3	This command will read a byte from EEPROM location 15 and place the result into variable b2, and then read a byte from EEPROM location 16 and place the result into variable b3.
write 0, b0	This command will write the byte in b0 to EEPROM location 0.
write 7, b2, b5	This command will write the byte in b2 to EEPROM location 7 and the byte in b5 to EEPROM location 8.

TABLE

The X1- and X2-series chips have the facility to reserve a space for table data. The table is a 256-byte area of read-only storage in program memory. The table is loaded at the time of program download and the data in it can be read during the execution of a program.

The commands that operate on table are:

table

readtable

The syntax of the table command is:

table {address,} (data {, data} ...)

where

address is an optional constant in the range 0–255 that specifies an address in the table.

data is a constant that specifies the data to be stored.

This is a compiler command and address and data must be specified as constants. The address is optional and, if not specified, defaults to the next available address, or 0 (zero), if no previous address or data has been specified.

The syntax of the readtable command is:

readtable address, data

where

address is a variable of constant in he range 0–255 that specifies the table address.

data is a variable that receives the data.

Example:

```
table ("ABCDEF")       'Write the letters "ABCDEF" to table addresses 0–5.
readtable 1, b0        'Read table address 1 ( contains "B") into b0.
```

POINTERS

A pointer is a variable that contains the address of another variable. Pointers can be used in commands in place of a variable. There are two pointers available to the PICAXE programmer: **ptr** and **bptr.** The X1 series supports **ptr** and the X2 series supports both **ptr** and **bptr.**

Scratchpad Pointer   The scratchpad pointer points to addresses in scratchpad memory. It is a number in the range 0 – 127 for the PICAXE-28X1, -40X1, and -20X2 chips, and in the range 0 to 1,023 for the PICAXE-28X2 and -40X2 chips. The bits of the scratchpad pointer may be addressed individually as ptr9, ptr8, ptr7, ptr6, ptr5, ptr4, ptr3, ptr2, ptr1, and ptr0, and the individual bytes can be addressed as **ptrh** and **ptrl,** where **ptrl** is the least significant byte and **ptrh** is the most significant byte.

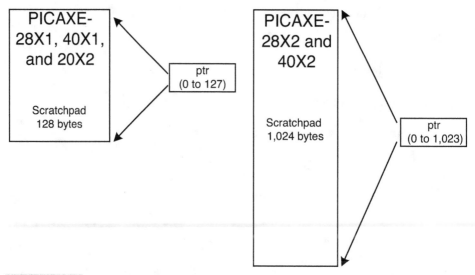

Figure 2.4 The scratchpad pointer.

The variable **ptr** is used in commands when setting the value of the scratchpad pointer. The symbol @ is placed in front of **ptr** to indicate that the variable to be used is the variable pointed to by **ptr** and not **ptr** itself. **@ptrinc** and **@ptrdec** can be used in the same places as **@ptr** to automatically increment or decrement **ptr** each time it is used. The increment or decrement takes place after the pointer is used. Figure 2.4 illustrates the scratchpad pointer.

Examples:

let ptr = 56	'Set the scratchpad pointer to point to address 56
let b0 = @ptr	'copy scratchpad address 56 to b0
let ptr = ptr + 1	'add 1 to the scratchpad pointer (now contains 57)
let @ptr = b0	'copy b0 to scratchpad address 57
ptr = 100	'Set the scratchpad pointer to point to address 100
b0 = @ptrinc	'copy scratchpad address 100 to b0 and increment the pointer
@ptr = b0	'copy b0 to scratchpad address 101
b0 = @ptrdec	'copy scratchpad address 101 to b0 and decrement the pointer
@ptrinc = b0	'copy b0 to scratchpad address 100 and increment the pointer

Byte-Scratchpad Pointer   The byte-scratchpad pointer points to addresses in the byte-scratchpad which is made up of byte variables and storage variables as illustrated in Fig. 2.5. It is a number in the range 0 to 255 and the bits may be addressed individually as bptr7, bptr6, bptr5, bptr4, bptr3, bptr2, bptr1, and bptr0.

All general-purpose variables can be addressed by using the byte-scratchpad pointer.

The variable **bptr** is used in commands when setting the value of the byte scratchpad pointer. The symbol @ is placed in front of **bptr** to indicate that the variable to be used

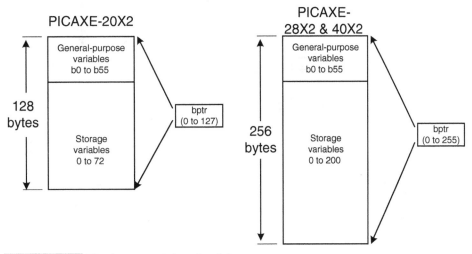

Figure 2.5 The byte-scratchpad pointer.

is the variable pointed to by **bptr** and not **bptr** itself. **@bptrinc** and **@bptrdec** can be used in the same places as **@bptr** to automatically increment or decrement **bptr** each time it is used. The increment or decrement takes place after the pointer is used.

Examples:

let bptr = 22   'Set the byte-scratchpad pointer to point to address 22 (which is
                        also b22)
let b0 = @bptr   'Copy byte-scratchpad address 22 (also b22) to b0
let bptr = bptr +1 'Add 1 to the byte-scratchpad pointer
let @bptr = b0   'Copy b0 to byte-scratchpad address 23 (also b23)

bptr = 120   'Set the byte-scratchpad pointer to point to address 120
b0 = @bptrinc   'Copy byte-scratchpad address 120 to b0 and increment the pointer
@bptr = b0   'Copy b0 to byte-scratchpad address 121
b0 = @bptrdec   'Copy byte-scratchpad address 121 to b0 and decrement the pointer
@bptrinc = b0   'Copy b0 to byte-scratchpad address 120 and increment the pointer

'This example will increment each of bytes 110–125 of the byte-scratchpad by 1
for bptr = 110 to 125   'Begin a for . . . next loop
inc @bptr   'Add 1 to the byte at the address pointed to by bptr
next bptr

Ports

All PICAXE chips have connections to the outside world called ports. Ports have external connections called legs (which are the electrical connection pins on the chip—not to be confused with pins, which are the internal references to the legs), and variables

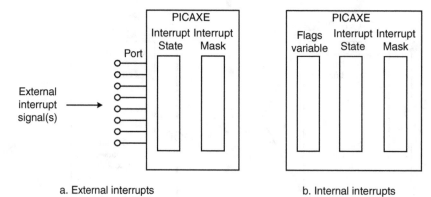

a. External interrupts   b. Internal interrupts

Figure 2.6 PICAXE interrupt structure. (a) External interrupts; (b) internal interrupts.

associated with them such as **pins, outpins,** and **dirs.** On PICAXE chips that have more than one port, the additional ports are referred to as **porta, portb,** etc., and the associated variables as **pinsa, pinsb,** etc. X2-series ports are addressed by a letter and a number separated by a dot. The letter identifies the port and the number the bit [e.g., C.0 refers to bit 0 (zero) of port C, B.4 refers to bit 4 of port B].

Ports are discussed further elsewhere in this book.

Interrupts

An interrupt is a means of informing the processor that an event has occurred that needs to be attended to. The event may be external, such as an input pin-changing state, or internal, such as a timer reaching a predetermined value. When interrupts are enabled and an interrupt event occurs, the processor will stop current processing and service the interrupt. When the interrupt has been serviced, the processor will resume where it left off when the interrupt occurred.

The PICAXE implements interrupts by means of polling and can be programmed to respond to interrupts that occur externally on input pins or internally in response to the flags variable. Figure 2.6 illustrates the interrupt structure.

"Interrupt state" and "interrupt mask" are internal values that are set using the **setint** and **setintflags** commands.

INTERFACING AND INPUT

OUTPUT TECHNIQUES

PICAXE Input and Output

All PICAXE chips have input and output pins that enable electrical signals to be received from, or sent to, external devices. The earlier PICAXE chips tend to dedicate pins to either input or output, with a few pins that can be configured as either. The more recent PICAXE chips allow most pins to be configured as input or output under program control.

PICAXE chips have facilities for:

- Digital input and output
- Analog input
- Asynchronous serial input and output
- I2C bus communication
- SPI bus communication
- UNI/O bus communication
- One-wire bus communication
- PC keyboard input
- Infrared input and output
- Pulse input and output
- Pulse-width modulation output
- Servo motor output
- Sound and music output

Hardware Interfacing

Logic Levels The digital inputs of standard voltage (4.2 to 5.5 V) PICAXE chips recognize +0.8 V or less as logic low and +2.0 V or more as logic high. The region between +0.8 V and +2.0 V is undefined. The digital outputs will guarantee +0.6 V or less for a logic low and 0.7 V less than the supply voltage, or more, for a logic high. These levels are compatible with TTL levels and other devices that operate at TTL levels can be connected to a standard voltage PICAXE input or output.

Low-voltage (1.8 to 3.6 V) PICAXE chips recognize 20% of the supply voltage, or less, as logic low and 0.8 V plus 25% of the supply voltage, or more, as logic high. The digital output will guarantee +0.6 V or less for a logic low and 0.7 V less than the supply voltage, or more, for a logic high. These levels, in particular logic high, may not be compatible with TTL levels.

Consideration must also be given to the voltage limits that can be applied to an input. The voltage on a PICAXE input must not be more than 0.3 V below ground (0 V) or greater than 0.3 V above the supply voltage. The application of voltages outside these limits may cause unpredictable results or possibly cause damage to the chip.

Digital logic levels are summarized in Table 3.1.

Converting between 5 and 3.3 V logic The voltage at a digital input of a PICAXE running with a 3.3 V supply must not exceed 3.6 V (supply 3.3 V+ 0.3 V) and if there is a need to interface 5-V logic to a PICAXE running at 3.3 V, then some form of level conversion is required. The simplest form that such a converter can take is a voltage divider that divides by two-thirds. Another way of converting 5-V logic to

TABLE 3.1   LOGIC LEVELS

STANDARD VOLTAGE (4.2 TO 5.5 V) CHIPS

Logic level	Voltage level for input	Voltage level for output
High	>= +2 V	>= Supply voltage − 0.7 V
Low	<= +0.8 V	<= +0.6 V

Low voltage (1.8 to 3.6 V) chips

Logic level	Voltage level for input	Voltage level for output
High	>= 25% of the supply voltage + 0.8 V	>= Supply voltage − 0.7 V
Low	<= 20% of the supply voltage	<= +0.6 V

3.3 V logic is to use a buffer chip that is designed for logic conversion. The 74LVX245, which has 5 V tolerant inputs, can perform this function. Figure 3.1 illustrates methods of 3.3/5 V conversion.

The output voltage level from a PICAXE running at 3.3 V falls within the range for TTL inouts and it is, therefore, possible to connect a 3.3 V logic output directly to a 5 V logic input.

Output current from a digital output pin   PICAXE output pins are capable of supplying currents up to 20 mA and will current-limit to about 25 mA. This makes them suitable for directly driving low current loads, such as, light-emitting diodes and small piezo speakers. Current-limiting resistors may not be strictly required in some applications, such as driving LEDs; however, it is a good idea to use them to reduce power dissipation within the chip.

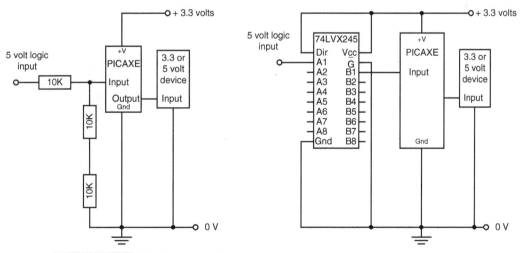

Figure 3.1   3.3/5 V conversion.

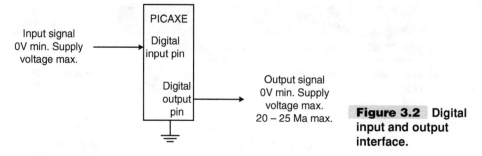

Figure 3.2 Digital input and output interface.

Digital Interfacing

Each PICAXE chip has input and output pins that can be used for digital interfacing; the electrical connections are shown in Fig 3.2.

The commands and variables associated with digital interfacing are:

high

low

input

output

if . . . then . . . endif

pins variable

outpins variable

dirs variable

The syntax of the **high, low, input,** and **output** commands is:

high pin {,pin . . . }

low pin {,pin . . . }

input pin {,pin . . . }

output pin {,pin . . . }

where

pin is a valid pin number.

Examples:

high 0

high 0, 1, 2

low 1

The **if** ... **then** ... **endif** command is a generic command for making decisions of all types and can be used to test the state of an input pin. The command is discussed further elsewhere in this book.

Example:

if pin3 = 1 then

    {commands}

endif

The variables **pins, dirs,** and **outpins** are associated with the PICAXE ports, although some ports may not have all three variables associated with them. Many PICAXE chips have several input/output ports and the additional ports are identified by a letter, e.g., porta, portb, portc. X2 chips identify ports by a letter followed by a number with a dot in between, e.g., A.0, B.2.

The variable **pins** will refer to the input pins when it is read (right-hand side of the assignment operator, e.g., b0 = pins) and will refer to the output pins when it is written (left-hand side of the assignment operator, e.g., pins = b0). The variable **outpins** can also be used to refer to the output pins and must be used when it is necessary to read the state of the output pins, e.g., b0 = outpins.

The **dirs** system variable sets the direction of the pins that can be configured as either input or output, a 1 (one) in a bit position of the **dirs** variable makes the corresponding pin an output pin. **dirs** is not implemented for ports that are dedicated as input or output ports. The **high, low, output,** and **input** commands can also be used to set the direction of configurable pins. Figure 3.3 shows the general arrangement for the **pins, dirs,** and **outpins** variables, and input/output ports.

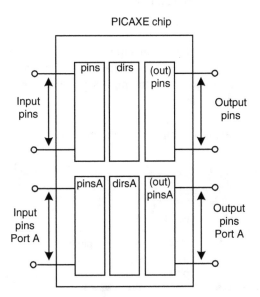

Figure 3.3 Relationship between the PICAXE pins and pins, dirs, and outpins variables.

Port Addressing

Digital output ports are addressed by their pin number. For earlier chips, this will be a number in the range 0 to 7, e.g., 0, 1, 2, and may be preceded in some commands by a port identifier, e.g., portc 1. For later chips, the port ID is included in front of the pin number in the form of a letter followed by a dot, e.g., A.0, A.1, B.0.

Digital input ports are addressed by the word "pin" followed by the pin number. For earlier chips, this will be a number in the range 0 to 7, e.g., pin0, pin1, pin2. For later chips, the port ID is included in front of the pin number in the form of a letter followed by a dot, e.g., pina.3, pinb.3, pina.0.

Digital inputs can also be read by reading the **pins** variable; digital outputs can be set by writing to the **pins** or **outpins** variable.

Setting the Direction of Configurable Pins

The direction of an input/output pin can be set by means of the **input** and **output** commands, by a **high** or **low** command referencing the pin, or by setting the corresponding bit in the **dirs** variable.

The PICAXE 08M has five digital I/O pins. Pin0 is a dedicated output pin, pin3 is a dedicated input pin, pin4, pin2, and pin1 can be configured as either inputs or outputs. Table 3.2 shows the layout of the **pins, outpins,** and **dirs** variables for the PICAXE-08M.

Examples for PICAXE-08M

high 1	'Set pin 1 as an output and place in the high state
low 1	'Set pin 1 as an output and place in the low state
input 1	'Make pin 1 an input pin
output 1	'Make pin 1 an output pin
pins = %00010111	'Set output pins high (I/O pins must be configured for output)
outpins = %00010111	'Set output pins high (I/O pins must be configured for output)
dirs = %00000010	'Make pin 1 an output pin and pin4 & pin2 input pins
if pin3 = 1 then {commands} endif	'Test the state of pin3
b0 = pins	'Read the state of the input pins and place into b0

TABLE 3.2 LAYOUT OF THE PINS, OUTPINS, AND DIRS VARIABLES FOR THE PICAXE-08M

PICAXE-08M	BIT 7	BIT 6	BIT 5	BIT 4	BIT 3	BIT 2	BIT 1	BIT 0
pins	-	-	-	pin4	pin3	pin2	pin1	-
outpins	-	-	-	pin4	-	pin2	pin1	pin0
dirs	-	-	-	pin4	-	pin2	pin1	-

TABLE 3.3 LAYOUT OF THE PINS, OUTPINS, AND DIRS VARIABLES FOR THE PICAXE-14M

PICAXE-14M	BIT 7	BIT 6	BIT 5	BIT 4	BIT 3	BIT 2	BIT 1	BIT 0
pins	-	-	-	pin4	pin3	pin2 (C5)	pin1 (C4)	pin0 (C3)
outpins	-	-	pin5 (C2)	pin4 (C1)	pin3 (C0)	pin2	pin1	pin0
portc		.	pin5	pin4	pin3	pin2	pin1	pin0
dirsc	-	-	portc pin5	portc pin4	portc pin3	portc pin2	portc pin1	portc pin0
dirsc initial value	x	x	0	0	0	1	1	1

The PICAXE-14M has 11 digital I/O pins, 2 are dedicated input pins, 3 are dedicated output pins, and 6 can be configured as either inputs or outputs. The pins are divided into two ports called pins and port C. The port C pins can be reconfigured and physically map some of the input and output pins. Table 3.3 shows the layout of the **pins, outpins,** and **dirs** variables for the PICAXE-14M.

Examples for PICAXE-14M

high 0	'Set pin 0 high
low 1	'Set pin 1 low
b0 = pins	'Read the state of the input pins and place into b0
pins = %00111111	'Set all output pins high (depends if pin is input or output)
outpins = %00111111	'Set all output pins high (depends if pin is input or output)
dirsc = %00111000	'Make pins 5,4, & 3 (C2,C1,C0) output, pins 2,1, & 0 (C5,C4,C3) input
if pin3 = 1 then {commands} endif	'Test the state of pin3
if portc pin1 = 0 then {commands} endif	'Test the state of C0

The PICAXE-18M and -18X have 13 digital I/O pins, 5 are dedicated as inputs, and 8 are dedicated as outputs. There are no pins that can be configured as either inputs or outputs. Table 3.4 shows the layout of the **pins** and **outpins** variables for the PICAXE-18M and -18X.

TABLE 3.4 LAYOUT OF THE PINS AND OUTPINS VARIABLES FOR THE PICAXE-18M AND -18X

PICAXE-18M/-18X	BIT 7	BIT 6	BIT 5	BIT 4	BIT 3	BIT 2	BIT 1	BIT 0
pins	pin7	pin6	-	-	-	pin2	pin1	pin0
outpins	pin7	pin6	pin5	pin4	pin3	pin2	pin1	pin0

TABLE 3.5 LAYOUT OF THE PINS AND OUTPINS VARIABLES FOR THE PICAXE-20M								
PICAXE-20M	**BIT 7**	**BIT 6**	**BIT 5**	**BIT 4**	**BIT 3**	**BIT 2**	**BIT 1**	**BIT 0**
pins	pin7	pin6	pin5	pin4	pin3	pin2	pin1	pin0
outpins	pin7	pin6	pin5	pin4	pin3	pin2	pin1	pin0

Examples for PICAXE-18M and -18X

```
high 1                 'Set pin 1 to the high state
low 1                  'Set pin 1 to in the low state
b0 = pins              'Read the state of the input pins and place into b0
pins = %11000111       'Set all output pins high (I/O pins must be configured for
                           output)
outpins = %11000111    'Set all output pins high (I/O pins must be configured for
                           output)
if pin2 = 1 then       'Test the state of pin2
   {commands}
endif                  'Test the state of pin2
```

The PICAXE-20M has 16 digital I/O pins, 8 are dedicated as inputs, and 8 are dedicated as outputs. Table 3.5 shows the layout of the **pins** and **outpins** variables for the PICAXE-20M.

Examples for PICAXE-20M

```
high 1                 'Set pin 1 to the high state
low 1                  'Set pin 1 to in the low state
b0 = pins              'Read the state of the input pins and place into b0
pins = %11000111       'Set all output pins high (depends if pin is input or output)
outpins = %11000111    'Set all output pins high (depends if pin is input or output)
if pin2 = 1 then       'Test the state of pin2
   {commands}
endif
```

The PICAXE-28X1 has 20 digital I/O pins, 8 are dedicated as outputs, 4 are dedicated as inputs, and 8 can be configured as either inputs or outputs. The pins are divided into three ports called port A, port B, and port C. Port A has 4 pins that are dedicated as inputs, port B has 8 pins that are dedicated as outputs, and port C has 8 pins that can be configured as either inputs or outputs. Table 3.6 shows the layout of the **pins, outpins,** and **dirs** variables for the PICAXE-28X1.

Examples for PICAXE-28X1 and -40X1

```
high 1                 'Set output pin 1 high
low 1                  'Set output pin 1 low
pins = %11000011       'Set output pins 7, 6, 1, & 0 high
b1 = outpins           'Read the state of the output pins into b1
```

TABLE 3.6 LAYOUT OF THE PINS, OUTPINS, AND DIRS VARIABLES FOR THE PICAXE-28X1

PICAXE-28X1	BIT 7	BIT 6	BIT 5	BIT 4	BIT 3	BIT 2	BIT 1	BIT 0
pins	pin7	pin6	pin5	pin4	pin3	pin2	pin1	pin0
pinsc	pin7	pin6	pin5	pin4	pin3	pin2	pin1	pin0
portc	pin7	pin6	pin5	pin4	pin3	pin2	pin1	pin0
dirsc	pin7	pin6	pin5	pin4	pin3	pin2	pin1	pin0
outpins (port B)	pin7	pin6	pin5	pin4	pin3	pin2	pin1	pin0
port A	-	-	-	-	pin3	pin2	pin1	pin0

if pin1 = 0 then endif   'Test the state of input pin 1
if porta pin0 = 1 then   'Test the state of port A pin 0
    endif
b0 = pins   'Read input pins into b0
dirsc = %11000011   'Set port c pins 7, 6, 1, & 0 as output pins
pinsc = %11000011   'Set port c pins 7, 6, 1, & 0 high (must be configured as outputs)
portc = %11000011   'Set port c pins 7, 6, 1, & 0 high (must be configured as outputs)

The PICAXE-40X1 has 28 digital I/O pins, 8 are dedicated as outputs, 12 are dedicated as inputs, and 8 can be configured as either inputs or outputs. The pins are divided into five ports called port A, port B, port C, port D, and port E. Port A has four pins that are dedicated as inputs, port B has eight pins that are dedicated as outputs, port C has eight pins that can be configured as either inputs or outputs, port D has eight pins that are dedicated as inputs, and port E does not have any digital pins. Table 3.7 shows the layout of the **pins, outpins,** and **dirs** variables for the PICAXE-40X1.

Examples for PICAXE-40X1 are the same as the examples for the PICAXE-28X1.

The PICAXE-20X2 has 17 digital I/O pins which can all be configured as either inputs or outputs. The pins for X2 series chips are addressed by the port Id and the

TABLE 3.7 LAYOUT OF THE PINS, OUTPINS, AND DIRS VARIABLES FOR THE PICAXE-40X1

PICAXE-40X1	BIT 7	BIT 6	BIT 5	BIT 4	BIT 3	BIT 2	BIT 1	BIT 0
pins (port D)	pin7	pin6	pin5	pin4	pin3	pin2	pin1	pin0
outpins (port B)	pin7	pin6	pin5	pin4	pin3	pin2	pin1	pin0
pinsa	-	-	-	-	pin3	pin2	pin1	pin0
pinsc	pin7	pin6	pin5	pin4	pin3	pin2	pin1	pin0
outpinsc	pin7	pin6	pin5	pin4	pin3	pin2	pin1	pin0
dirsc	pin7	pin6	pin5	pin4	pin3	pin2	pin1	pin0

TABLE 3.8 LAYOUT OF THE PINS AND DIRS VARIABLES FOR THE PICAXE-20X2

PICAXE-20X2	BIT 7	BIT 6	BIT 5	BIT 4	BIT 3	BIT 2	BIT 1	BIT 0
pinsA	-	-	-	-	-	-	-	A.0
dirsA	-	-	-	-	-	-	-	A.0
pinsB	B.7	B.6	B.5	B.4	B.3	B.2	B.1	B.0
dirsB	B.7	B.6	B.5	B.4	B.3	B.2	B.1	B.0
pinsC	C.7	C.6	C.5	C.4	C.3	C.2	C.1	C.0
dirsC	C.7	C.6	C.5	C.4	C.3	C.2	C.1	C.0

pin number separated by a dot (see examples below). Table 3.8 shows the layout of the **pins, outpins,** and **dirs** variables for the PICAXE-20X2.

The PICAXE-28X2 has 21 digital I/O pins which can all be configured as either inputs or outputs. The pins for X2 series chips are addressed by the port Id and the pin number separated by a dot (see examples below). Table 3.9 shows the layout of the **pins, outpins,** and **dirs** variables for the PICAXE-28X2.

The PICAXE-40X2 has 32 digital I/O pins which can all be configured as either inputs or outputs. The pins for X2 series chips are addressed by the port Id and the pin number separated by a dot (see examples below). Table 3.10 shows the layout of the **pins, outpins,** and **dirs** variables for the PICAXE-40X2.

Examples for PICAXE-20X2, -28X2, and -40X2

```
high A.0                  'Set port A pin 0 high
low B.1                   'Set port B pin 1 low
pinsb = %11000011         'Set port B pins 7, 6, 1, & 0 high
b1 = pinsb                'Read the state of port B into b1
b2 = outpinsb             'Read the state of port B output pins into b2
if pinB.1 = 0 then endif  'Test the state of port B pin 1
if pinB.0 = 1 then endif  'Test the state of port B pin 0
dirsb = %11000011         'Set port C pins 7, 6, 1, & 0 as output pins
pinsb = %11000011         'Set port C pins 7, 6, 1, & 0 high (must be configured as
                           outputs)
```

TABLE 3.9 LAYOUT OF THE PINS AND DIRS VARIABLES FOR THE PICAXE-28X2

PICAXE-28X2	BIT 7	BIT 6	BIT 5	BIT 4	BIT 3	BIT 2	BIT 1	BIT 0
pinsA	-	-	-	A.4	A.3	A.2	A.1	A.0
dirsA	-	-	-	A.4	A.3	A.2	A.1	A.0
pinsB	B.7	B.6	B.5	B.4	B.3	B.2	B.1	B.0
dirsB	B.7	B.6	B.5	B.4	B.3	B.2	B.1	B.0
pinsC	C.7	C.6	C.5	C.4	C.3	C.2	C.1	C.0
dirsC	C.7	C.6	C.5	C.4	C.3	C.2	C.1	C.0

TABLE 3.10   LAYOUT OF THE PINS AND DIRS VARIABLES FOR THE PICAXE-40X2

PICAXE-40X2	BIT 7	BIT 6	BIT 5	BIT 4	BIT 3	BIT 2	BIT 1	BIT 0
pinsA	A.7	A.6	A.5	A.4	A.3	A.2	A.1	A.0
dirsA	A.7	A.6	A.5	A.4	A.3	A.2	A.1	A.0
pinsB	B.7	B.6	B.5	B.4	B.3	B.2	B.1	B.0
dirsB	B.7	B.6	B.5	B.4	B.3	B.2	B.1	B.0
pinsC	C.7	C.6	C.5	C.4	C.3	C.2	C.1	C.0
dirsC	C.7	C.6	C.5	C.4	C.3	C.2	C.1	C.0
pinsD	D.7	D.6	D.5	D.4	D.3	D.2	D.1	D.0
dirsD	D.7	D.6	D.5	D.4	D.3	D.2	D.1	D.0

Analog Interfacing

Each PICAXE chip has several pins that can be used for 8 bit or 10 bit analog input. The electrical connections are shown in Fig 3.4.

The commands and variable associated with analog input are:

readadc

readadc10

calibadc

calibadc10

adcsetup variable

The **readadc** command will read an analog input with 8-bit resolution and the **readadc10** command has 10-bit resolution. Analog readings are normally referenced to the PICAXE supply voltage and, if the supply voltage varies with battery discharge or

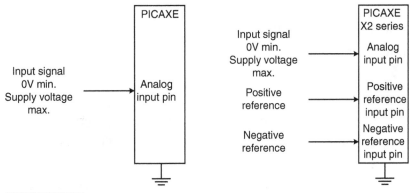

Figure 3.4   Electrical connections for analog input.

TABLE 3.11   SUMMARY OF ANALOG INPUT CHARACTERISTICS

PICAXE	ANALOG CHANNELS	INTERNAL VOLTAGE REFERENCE	EXTERNAL NEGATIVE REFERENCE	EXTERNAL POSITIVE REFERENCE	VALID PINS/CHANNELS
08M	3				1, 2, 4
14M	2 + 3				0, 4 + 1, 2, 3
18M	3				0, 1, 2
20M	4	0.6 V			1, 2, 3, 7
18X	3				0, 1, 2
28X1	4				0 to 3
40X1	7				0 to 3, & 5 to 7
20X2	11	1.024 Volts	ADC5	ADC4	Channel 1 to 11
28X2	9	1.2 Volts— low voltage parts only	ADC2	ADC3	Channel 0 to 3, & 8 to 12
40X2	12	1.2 Volts— low voltage parts only	ADC2	ADC3	Channel 0 to 3, & 5 to 12

regulator tolerance, for example, then the analog reading will also vary. The PICAXE-20M, -20X2, -28X1, -40X1, -28X2-3V, and -40X2-3V chips have an internal voltage reference that can be read by the **calibadc** and **calibadc10** commands; the X2 series chips can also be configured to use an external reference source.

The **calibadc** and **calibadc10** commands read a calibration value with 8- or 10-bit resolution, this value can then be used to compensate for variations in supply voltage.

For X2 parts, it is necessary to configure analog input pins before they can be used and the **adcsetup** variable is provided for this purpose. The **adcsetup** variable can also be used to configure an external analog reference source.

Analog ports are addressed by pin number for M-, X-, and X1-series chips and by channel number for X2-series chips. Analog input characteristics are summarized in Table 3.11.

The syntax of the **readadc** and **readadc10** commands is

readadc *channel*, *variable*

readadc10 *channel*, word*variable*

where

channel is a variable or constant that specifies the analog port.

variable is a variable that receives the reading and must be a word for **readadc10**

Note that Revolution Education recommends a modified serial programming circuit when **readadc10** and **debug** commands are used in the same program.

TABLE 3.12   LAYOUT OF THE ADCSETUP VARIABLE

BIT	USAGE 20X2, 28X2-3V, 40X2-3V	USAGE 28X2 (5 V PARTS ONLY)	USAGE 40X2 (5 V PARTS ONLY)
0	ADC 0, 1= configured	0 = None	0 = None
1	ADC 1, 1= configured	1 = ADC0	1 = ADC0
2	ADC 2, 1= configured	1 = ADC0,1	1 = ADC0,1
3	ADC 3, 1= configured	1 = ADC0,1,2	1 = ADC0,1,2
4	ADC 4, 1= configured	1 = ADC0,1,2,3	1 = ADC0,1,2,3
5	ADC 5, 1= configured	1 = ADC0,1,2,3,8	1 = ADC0,1,2,3,5
6	ADC 6, 1= configured	1 = ADC0,1,2,3,8,9	1 = ADC0,1,2,3,5,6
7	ADC 7, 1= configured	1 = ADC0,1,2,3,8,9,10	1 = ADC0,1,2,3,5,6,7
8	ADC 8, 1= configured	1 = ADC0,1,2,3,8,9,10,11	1 = ADC0,1,2,3,5,6,7,8
9	ADC 9, 1= configured	1 = ADC0,1,2,3,8,9,10,11,12	1 = ADC0,1,2,3,5,6,7,8,9
10	ADC 10, 1= configured		1 = ADC0,1,2,3,5,6,7,8,9,10
11	ADC 11, 1= configured		1 = ADC0,1,2,3,5,6,7,8,9,10,11
12	ADC 12, 1= configured		1 = ADC0,1,2,3,5,6,7,8,9,10,11,12
13	Not used		
14	External positive reference	1 ADC3 = +ve ref. for 28X2 & 40X2, ADC4 = +ve ref. for 20X2 0 = Internal +ve reference equal to supply voltage	
15	External negative reference	1 ADC2 = −ve ref. for 28X2 & 40X2, ADC5 = −ve ref. for 20X2 0 = Internal −ve reference equal to 0 (zero) V	

The syntax of the **calibadc** and **calibadc10** commands is

calibadc *variable*

calibadc10 *variable*

where

variable is a variable that receives the calibration reading and must be a word for **calibadc10**

adcsetup is a word variable, the layout is given in the PICAXE manual and is reproduced in Table 3.12 for convenience.

Bits 14 and 15 of **adcsetup** can be used to enable external voltage references for the X2-series chips. When external voltage references are enabled, the corresponding pins will be connected to external reference voltages.

Bits 0 – 12 of **adcsetup** configure the analog ports for use and there are different settings for the 20X2, 28/40X2-3V, and 28/40X2 5 Volt parts. For the 20X2, 28/40X2-3V, a bit in any position will configure the corresponding analog port. The 28/40X2 5 V parts use an incremental scheme for configuring analog ports and to configure a particular analog port, all lower numbered analog ports must also be configured. For example, to configure analog port 3 on the 28/40X2 5 V parts, analog ports 2, 1, and 0 must also be configured as analog ports. Note also that analog ports 4, 5, 6, and 7 are not implemented on the 28X2 5 V chip, and analog port 4 is not implemented on the 40X2 5 V chip.

Parallel and Serial Interfacing

Microcontrollers can perform many tasks by themselves, however, it is becoming increasingly common to transmit data between microcontrollers and other devices, such as, sensors, memory devices, and computers. Transmitting data requires an interface, which is simply a point of interaction between two devices. The common unit of transmission is the byte, although other units can be used.

A parallel interface sends all the bits of a byte at the same time, which requires a single connection for each bit; these connections are called data lines. There is also a control line to indicate that data is present on the data lines. Parallel interfaces are impractical for long-distance communications and even for short distances they are becoming less common in favor of serial interfaces. A serial interface transmits data bits one after the other on a single data line and a number of different protocols are in use. The protocols may be implemented in software or hardware. Software implementations use more processor resources and take up more program space. Hardware implementations tend to be preferred where there is a need to reduce processor load and software complexity. Figure 3.5 shows a comparison between an 8-bit parallel interface and an asynchronous serial interface.

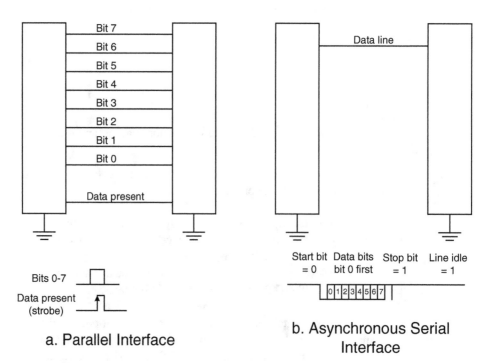

Figure 3.5   Parallel (a) and serial Interfaces (b).

TABLE 3.13 SERIAL INTERFACES SUPPORTED BY PICAXE		
PROTOCOL	**ABBREVIATION**	**NUMBER OF WIRES**
Asynchronous RS232	Asynch or RS232	1 data line per direction
Inter Integrated Circuit	I2C or I²C	1 data + 1 clock
Serial Peripheral Interface	SPI	2 data + clock + device select
1-wire	1-wire	1 data + 1 power (optional)
UNI/O	UNI/O	1 data + 1 power (optional)

Perhaps the best known serial protocol is the asynchronous serial protocol commonly called RS232 (although RS232 actually defines voltage levels and other aspects of the signal path rather than the protocol). The asynchronous serial protocol is commonly used to transmit bytes between two devices and the code most often used is the American Standard Code for Information Interchange, abbreviated ASCII. Other serial transmission protocols are in use; they may use separate lines for data and for clock pulses. Table 3.13 summarizes the serial protocols supported by the PICAXE.

Asynchronous–RS232

Asynchronous serial communication can use one or two wires. Transmission in a single direction only, called simplex transmission, needs only one wire. Transmission in both directions can be half-duplex, which uses one wire, or full-duplex, which requires two wires. When using half-duplex, data can be transmitted in both directions but only one direction at a time. Full-duplex transmission allows data to be transmitted simultaneously in both directions and is effectively two simplex links. Figure 3.6 shows the communication link types.

The transmission levels are typically TTL or RS232, the output of a PICAXE is TTL level, which is suitable for short distances. For longer distances, RS232 levels are normally used. Table 3.14 summarizes the levels and states.

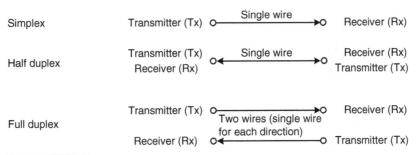

Figure 3.6 Communication link types.

TABLE 3.14   COMPARISON OF TTL AND RS232 LEVELS

STATES	BINARY	TTL	RS232
High	1	$>= 2$ V	-3 to -12 V
Low	0	$<= 0.5$ V	$+3$ to $+12$ V

The PICAXE supports asynchronous serial protocol in both software and hardware implementations; the associated commands are:

serin

serout

serrxd

sertxd

hsersetup

hserin

hserout

Electrical connections for RS232 asynchronous input and output are shown in Fig. 3.7.

The **serin** command will receive asynchronous serial data at a specified pin and baud rate; the **serout** command will transmit asynchronous serial data from a specified pin at a specified baud rate. The **serrxd** command will receive asynchronous serial data from the serial programming port input pin at a baud rate that is fixed relative to clock speed. The **sertxd** command will transmit asynchronous serial data to the serial programming port output pin at a baud rate that is fixed relative to clock speed.

The **hsersetup** command is used to setup the hardware serial port on the PICAXE for use by the **hserin** and **hserout** commands. The **hserin** command will receive asynchronous serial data at the hardware input pin at a baud rate that was specified in the previously executed **hsersetup** command. The **hserout** command will transmit asynchronous serial data from the hardware output pin at a baud rate that was specified in the previously executed **hsersetup** command.

All asynchronous serial transmission is performed with no parity, 8 data bits, and 1 stop bit (N, 8, 1).

The syntax of the **serin** command is:

serin {[timeout, label],} port, mode {,(qualifier)} {, (qualifier) ... } {, {#} variable} {,{#} variable ... }

where

port is a variable or constant which specifies the serial input port.

mode is a constant or variable that specifies one of the transmission modes.

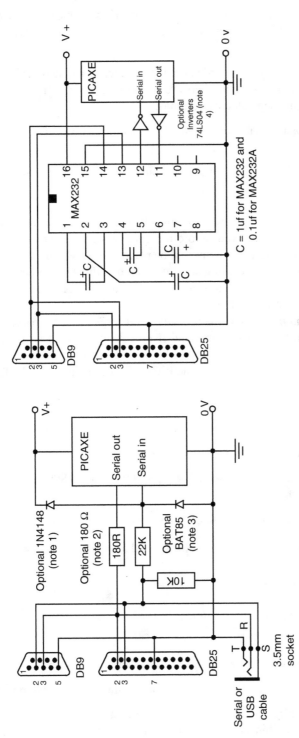

Note 1. Required for 08M & 14M input 3, and 20X2 input c.6.
Note 2. For short-circuit protection only, can be omitted for programming circuit and most applications.
Note 3. Required where readadc10 and debug commands are used in a program.

a. Serial port connection

Note 4. Required only where inverted levels are used (such as programming circuit), not required where true levels are used.
Note 5. RS232 levels should not be applied to the USB Programming cable (AXE027)

b. Serial port connection with true RS232 conversion

Figure 3.7 Electrical connections for asynchronous serial protocol. (a) Serial port connection; (b) serial port connection with true RS232 conversion.

qualifier is an optional constant or variable that specifies a value that must be received before following bytes are stored. Qualifiers must be enclosed in brackets and there may be more than one.

variable is an optional byte or word variable that receives a data byte; if preceded by the "#" character, then ASCII numbers are converted to binary. If there is more than one variable, they are separated by commas.

timeout is an optional variable or constant that specifies a timeout period in milliseconds.

label is a label in the program where program execution will resume if a timeout occurs.

Qualifiers are optional and, if specified, must be enclosed in brackets; the qualifiers themselves and any bytes received before the qualifiers are discarded. Variables are optional and, if specified, received bytes will be placed consecutively into the variables specified in the command. Processing stops until all qualifiers have been received and all variables have been filled, or until a timeout occurs. There must be at least one qualifier or variable specified in the command.

The *timeout* and *address* keywords are only valid for X1 and X2 parts and provide a means of exiting from a **serin** command if no data is received within a specified time.

The transmission modes are listed in the PICAXE manual and take the form:

Lnnnn_cc

where

L is "T" for true levels or "N" for inverted levels.

nnnn is the baud rate and may be 300, 600, 1,200, 2,400, 4,800, 9,600, or 19,200.
_ is an underscore character.

cc is the clock speed and may be 4, 8, or 16.

Some combinations of baud rate and clock speed are not valid.

The syntax of the **serout** is:

serout *port, mode,* ({#}*data* {,{#}*data*} ...)

where

port is a constant or variable which specifies the serial output port.

mode is a constant or variable that specifies one of the transmission modes.

data is a constant or byte variable that contains the data to be written; if preceded by the "#" character, then binary numbers are converted to ASCII. There may be more than one constant or variable and constants and variables may be mixed.

Examples:

serin 1, T2400_4, b0

Pause program execution until a byte is received at 2,400 bauds, no parity, 8-data bits, and 1 stop bit from serial input port 1 when running at a clock speed of 4 MHz and true transmission level. The command stores the received byte in variable b0 and program execution continues.

serin [30, label1], 1, T2400_4, b0

This code, for X1- and X2-series chips, pauses program execution until a byte is received at 2,400 bauds, no parity, 8-data bits, and 1 stop bit from serial input port 1 when running at a clock speed of 4 MHz and true transmission level. The command stores the received byte in variable b0 and program execution continues. If a byte is not received within 30 ms, program execution resumes at *label1*.

b0 = 3

b1 = N1200_4

serin b0, b1, ("AB"), b2, b3, b4

Pause program execution until the character "A" followed by "B" is received at 1,200 bauds, no parity, 8-data bits, and 1 stop bit from serial input port 3 when running at a clock speed of 4 MHz and inverted transmission level. The "A" and "B" are discarded along with any bytes that are received before the "A" and "B." The next three bytes received are stored in variables b2, b3, and b4 and program execution continues.

serout 1, T2400_4, (b0)

Transmit the byte in b0 to serial output port 1 at 2,400 bauds, with no parity, 8-data bits, and 1 stop bit at a clock speed of 4 MHz and true transmission level.

b1 = N1200_4

serout b0, b1, (b2, b3, b4)

Transmit the bytes in b2, b3, and b4 to the serial output port specified in b0, at 1,200 baud with no parity, 8-data bits, 1 stop bit, at a clock speed of 4 MHz and inverted transmission level.

The syntax of the **serrxd** command is:

serrxd {[timeout, label]} {,(qualifier)} {, (qualifier)} {, {#} variable} {, {#} variable} . . .

where

> *qualifier* is an optional constant or variable that specifies a value that must received before following bytes are stored. Qualifiers must be enclosed in brackets and there may be more than one.

> *variable* is an optional byte or word variable that receives a data byte; if preceded by the "#" character, then ASCII numbers are converted to binary. If there is more than one variable they are separated by commas.

> *timeout* is an optional variable or constant that specifies a timeout period in milliseconds.

> *label* is a label in the program where program execution will resume if a timeout occurs.

Qualifiers are optional and, if specified, must be enclosed in brackets; the qualifiers themselves and any bytes received before the qualifiers are discarded. Variables are optional and, if specified, received bytes will be placed consecutively into the variables specified in the command. Processing stops until all qualifiers have been received and all variables have been filled, or until a timeout occurs. There must be at least one qualifier or variable specified in the command.

The *timeout* and *address* keywords are only valid for X1 and X2 parts and provide a means of exiting from a **serrxd** command if no data is received within a specified time.

The transmission mode, for 4 MHz clock speed, is fixed at 4,800 bauds, no parity, 8-data bits, 1 stop bit, and inverted transmission level. The baud rate will increase proportionally with clock speed and baud rates of 9,600, 19,200, and 38,400 are possible.

Note that the **serrxd** command disables the internal scan for a new program being downloaded and it will not be possible to download a new program until a **reconnect** command is issued or the chip is reset. (See the **disconnect** and **reconnect** commands for more information.)

Examples:

serrxd b0

> Pause program execution until a byte is received at 4,800 bauds, no parity, 8-data bits, 1 stop bit, from the serial programming port when running at a clock speed of 4 MHz and inverted transmission level. The command stores the received byte in variable b0 and program execution continues.

The syntax of the **sertxd** is:

sertxd ({#}*data*{, {#}*data*} ...)

where

> *data* is a constant or byte variable that contains the data to be written; if preceded by a # character, then binary is converted to ASCII.

The transmission mode is fixed at 4,800 baud, no parity, 8-data bits, 1 stop bit, and inverted transmission level when running at a clock speed of 4 MHz. The baud rate will increase proportionally with clock speed and baud rates of 9,600 and 19,200 are possible. Example:

sertxd (b0, b1)

Send the bytes in b0 and b1 to the serial programming output port at 4,800 bauds, with no parity, 8-data bits, and 1 stop bit, and inverted transmission level when running at a clock speed of 4 MHz.

The syntax of the **hsersetup** command is

hsersetup off

hsersetup baudrate, mode

where

off is a keyword that turns the hardware serial port off.

baudrate is a variable or constant that specifies the baud rate.

mode is a variable or constant that specifies receive and transmit modes.

The **hserptr** and **hserinflag** variables are associated with hardware serial input. The baud rates are listed in the PICAXE manual and take the form:

Bnnnn_cc

where

B is the letter "B"

nnnn is the baud rate and may be 300, 600, 1,200, 2,400, 4,800, 9,600, 19,200, 38,400, 57,600, or 115,200.

_ is an underscore character.

cc is the clock speed and may be 4, 8, 16, 20, 32 or 40.

receive and transmit modes are listed in the PICAXE manual and are reproduced in Table 3.15 for convenience.

Example:

hsersetup off

Turn the hardware serial port off

hsersetup B9600_8, %001

TABLE 3.15 HARDWARE SERIAL MODES			
BITS 7 - 3	**BIT 2**	**BIT 1**	**BIT 0**
xxxxx	0 = input level True 1 = input level Inverted (X2 parts only)	0 = output level True 1 = output level Inverted	0 = no background receive 1= background receive into scratchpad

Configure the hardware serial port for background receive into scratchpad memory at 9,600 bauds, no parity, 8-data bits, 1 stop bit, true output level, true input level at a clock speed of 8 MHz. The **hserptr** variable is reset to zero and **hserinflag** is cleared.

Serial data will be received into the scratchpad beginning with the address in **hserptr** (which is zero), **hserptr** will be incremented for each byte received and the **hserinflag** flag will be set when a byte is received. The user program must test **hserinflag**, process the received data in the scratchpad, and reset **hserptr** and **hserinflag.**

The syntax of the **hserin** command is

    hserin {[timeout, label],} addr, count {, qualifier)}

where

    addr is a variable or constant specifying an address in scratchpad memory where the first received byte will be stored, *addr* will be incremented for each byte stored in scratchpad memory.

    count is a variable or constant specifying the number of bytes to be read from the serial port and stored in scratchpad memory.

    qualifier is an optional 8-bit constant or byte variable that specifies a value that must received before following bytes are stored.

    timeout is an optional variable or constant that specifies a timeout period in milliseconds.

    label is a label in the program where program execution will resume if a timeout occurs.

Examples:

hserin 0, 2

Read 2 bytes from the serial port and store the first byte in location 0 of scratchpad memory and the second byte in location 1 of scratchpad memory.

b0 = 5

b1 = 4

hserin b0, b1

Read 4 bytes from the serial port and store the first byte in location 5 of scratchpad memory; the following bytes will be stored in locations 6, 7, and 8.

hserin 0, 2, (10)

Read bytes from the serial port until a byte containing the qualifier 10 is received. Then, 2 more bytes will be read from the serial port. The first will be stored in location 0 of scratchpad memory and the second in location 1 of scratchpad memory.

The syntax of the **hserout** command is

hserout break, ({#} data {,{#} data} ...)

where

break is a variable or constant that specifies whether a break character should be sent; 0 = no break character, 1 = send break character.

data is a constant or byte variable that contains the data to be written; if preceded by a # character, then binary is converted to ASCII.

Examples:

hserout 0, (65)

Send a single byte containing the number 65 to the serial port. No break character is sent because the value for break is 0.

hserout 1, (b0, b1, b2)

Send a break character, then send bytes b0, b1, and b2 to the serial port.

Personal Computer Connectivity

The PICAXE can be connected to PCs, or other devices, using asynchronous serial protocol. Figure 3.7 shows the connections. In practice, the asynchronous serial interface uses common standards and almost any device meeting those standards can be connected to any other device that meets those standards.

Note that some devices implement serial communication at TTL levels whereas other devices use the RS232 levels of +/− 10 or 12 V and it may be necessary to convert between these two levels, as shown in Fig. 3.7b. A connection can be made between the PICAXE and a personal computer using the serial programming cable (part AXE026) using TTL or RS232 levels. The USB programming cable driver for the emulates a serial port and the USB cable can be used for a serial connection at TTL levels; RS232 levels should not be used with the USB programming cable. Connecting +/− 10 or 12-V levels to a device that is designed for 5 VTTL levels, such as the AXE027 USB programming cable, can cause permanent damage.

When communicating between a PICAXE and a PC it is necessary to use a program to receive the communication from the PICAXE. Any terminal program should be suitable, including Programming Editor terminal. Terminal programs allow data to be saved in a file and most spreadsheet and database software will accept data in csv (comma separated values) format. It is a simple matter to send data in csv format from a PICAXE. The csv file format consists of numeric or alphanumeric values in ASCII code separated by commas with a carriage return ($0D) and linefeed character ($0A) at the end of each line. For example, the number 123 and the letters "ABC" would be sent as the ASCII characters 123,ABC(CR)(LF). Some implementations of the csv format enclose values between double quotes (") and accept a (CR)(LF) or (LF)(CR) combination at the end of a line.

I2C

The I2C bus is a two-wire (plus ground) bus that carries data on one wire and clock pulses on the other. External pull-up resistors are required for both wires. Clock pulses are controlled by the master device that is active at the time of a data transfer and data travels in both directions on the data wire. Slave addresses can be assigned under software control (if the device supports it), be assigned at the time of manufacture, or be configured by pins on the device.

The I2C protocol supports multiple master devices and multiple slave devices. Only one master can be active at any one time and master devices must use appropriate handshaking to ensure that only one master device is active at any one time. Each slave device on the bus must have a unique address and the addresses consist of seven bits that are sent in the high-order bits of a byte. Sixteen of the possible 128 addresses are reserved.

Electrical connections for the I2C bus are shown in Fig. 3.8.

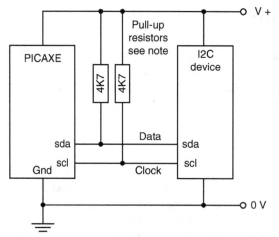

Note. Pull-up resistors are required for proper operation. Only one set per bus. Typical values are 1.8 K to 10 K. 4.7 K is a nominal value, longer buses and more devices require lower values

Figure 3.8 I2C **Electrical connections for the I2C bus.**

The PICAXE implementation of I2C supports a single master and multiple slave devices. Multiple master devices are possible with appropriate handshaking to ensure that only one master device is active at any one time.

The commands associated with I2C are:

i2cslave (slavei2c)

readi2c (i2cread)

writei2c (i2cwrite)

hi2csetup

hi2cin

hi2cout

The **i2cslave** command configures the PICAXE to use an I2C device at a specific I2C address, bus speed (100 or 400 KHz), and address length (byte or word). The **readi2c** command reads from the previously configured I2C device, and the **writei2c** command writes to the previously configured I2C device.

The **hi2csetup** command is used to setup the hardware I2C port on the PICAXE for use by the **hi2cin** and **hi2cout** commands. It also configures the PICAXE as either a master or slave device. The **hi2cin** command will read data from the hardware I2C port and the **hi2cout** command will write data to the hardware I2C port.

The syntax of the **i2cslave** command is:

i2cslave address, speed, addresslength

slavei2c address, speed, addresslength

where

address is the I2C device address.

speed is either "i2cslow" for devices that operate at 100 KHz or "i2cfast" for devices that can operate at 400 kHz. These constants have different syntax for increased clock speeds (see below).

addresslength is either "i2cbyte" for devices that have 8-bit addresses or "i2cword" for devices that have 16-bit addresses.

The *speed* constants for increased clock speeds are given in the PICAXE manual and are reproduced here for convenience:

i2cslow 100 kHz device at PICAXE clock of 4 MHz

i2cslow8 100 kHz device at PICAXE clock of 8 MHz

i2cslow16 100 kHz device at PICAXE clock of 16 MHz

i2cfast 400 kHz device at PICAXE clock of 4 MHz

i2cfast8 400 kHz device at PICAXE clock of 8 MHz

i2cfast16 400 kHz device at PICAXE clock of 16 MHz

The syntax of the **readi2c** command is:

readi2c {*address,*} (*data* {*,data* } ...)

i2cread {*address,*} (*data* {*,data* } ...)

where

address is an optional variable or constant specifying the address of the first I2C device register to be read. If not present, the next address is used, or address 0 if no prior address has been specified. *Address* will be either a byte or word, depending on the address length specified in the previous i2cslave command.

data is a variable that receives a data byte. If there is more than one variable, they are separated by commas.

The syntax of the **writei2c** command is:

writei2c {*address*}, (*data* {*,data*} ...)

i2cwrite {*address*}, (*data* {*,data*} ...)

where

address is a variable or constant specifying the byte or word address of the first I2C device register to be written.

data is a constant or byte variable that contains the data to be written.

Examples:

i2cslave %10100000, i2cfast, i2cword 'Initialize i2c for device address "1010000", 400 KHz, and word addressing.

readi2c 0, (b0)   'Read a byte into b0 from address 0
readi2c (b1, b2)   'Read a byte into b1 from addr 1, and a byte into b2 from addr 2

writei2c 0, (b0)   'Write a byte from b0 to address 0
writei2c (b1, b2)   'Write a byte from b1 to addr 1, and a byte from b2 to addr 2

The syntax of the **hi2csetup** command is:

hi2csetup off

hi2csetup i2cslave, slaveaddress

hi2csetup i2cmaster, slaveaddress, speed, addresslength

where

The *off* keyword turns the hardware i2c module off.

The *i2cslave* keyword sets the PICAXE to slave mode. This keyword must be followed by the *slaveaddress* that will be assigned to the PICAXE chip that is acting as a slave device.

The *i2cmaster* keyword sets the PICAXE to master mode,. This keyword must be followed by *slaveaddress*, *speed*, and *addresslength* of the slave device that is being addressed.

speed is either "i2cslow" for devices that operate at 100 KHz or "i2cfast" for devices that can operate at 400 kHz. These constants have different syntax for increased clock speeds (see below).

addresslength is either "i2cbyte" for devices that have 8-bit addresses or "i2cword" for devices that have 16-bit addresses.

The slave address is a 7-bit binary number, which must be placed in the high-order bits of a byte. Slave addresses can be found in the data sheets for the I2C device that is being used and may be split into two parts for some devices. For example, the 24LC256 EEP-ROM chip has the high-order 4 bits of its address fixed at the time of manufacture and the low-order 3 bits of the address are connected to pins on the chip. This allows up to eight different 24LC256 chips to be placed on a single bus without causing address conflicts.

Where a PICAXE has been configured as a slave device, any slave address can be assigned, although it should be noted that some addresses, beginning with either %1111xxxx or $0000xxxx, are reserved and should not be used. When a PICAXE is configured as a slave device it appears as a 256-byte memory chip and the memory is physically located in the scratchpad. The **hi2clast** and **hi2cflag** variables are associated with this mode of operation. A PICAXE configured as an I2C slave has its processing power available and can read from and write to scratchpad memory. Thus, for example, it could be programmed to read data from a sensor and place the results in scratchpad to be read by another PICAXE acting as an I2C master.

The PICAXE manual lists I2C device addresses; some of them are reproduced here for convenience:

24LC256	32K EEPROM	%1010dddx, i2cfast, i2cword
24LC512	64K EEPROM	%1010dddx, i2cfast, i2cword
DS1307	Real time clock	%1101000x, i2cslow, i2cbyte
AXE033	Serial LCD	%11001010, i2cslow, i2cbyte

The *speed* constants for increased clock speeds are given in the PICAXE manual and are reproduced here for convenience:

i2cslow 100-kHz device at 4 MHz

i2cslow_8 100-kHz device at 8 MHz

i2cslow_16 100-kHz device at 16 MHz

i2cfast 400-kHz device at 4 MHz

i2cfast_8 400-kHz device at 8 MHz

i2cfast_16 400-kHz device at 16 MHz

The syntax of the **hi2cin** command is:

hi2cin {*address,*} (*data* {, *data*} ...)

hi2cin [*newslave*], {*address*}, (*data* {,*data*} ...)

where

address is a variable or constant specifying the byte or word address of the first I2C device register to be read. *address* is optional and if left out, the address used will be the next address or address 0 if no previous **hi2cin** commands have been issued.

data is the PICAXE variable(s) that receives the data being read. There may be more than one variable and the number of bytes read from the device will correspond to the number of variables specified in the command.

newslave is an optional new slave address. It enables the slave address to be changed without the need to issue an **hi2csetup** command. The new slave address is applied to future **hi2cin** and **hi2cout** commands until it is changed by a subsequent **hi2csetup,** **hi2cin,** or **hi2cout** command.

The syntax of the **hi2cout** command is:

hi2cout {*address,*} (*data* {, *data*} ...)

hi2cout [*newslave*], {*address*}, (*data* {,*data*} ...)

where

address is a variable or constant specifying the byte or word address of the first I2C device register to be written. *address* is optional and if left out, the address used will be the next address or address 0 if no previous **hi2cout** commands have been issued.

data is the PICAXE variable(s) or constant(s) that are written to the device. There may be more than one variable or constant and the number of bytes written to the device will correspond to the number of variables and/or constants specified in the command.

newslave is an optional new slave address. It enables the slave address to be changed without the need to issue an **hi2csetup** command. The new slave address is applied to future **hi2cin** and **hi2cout** commands until it is changed by a subsequent **hi2csetup,** **hi2cin,** or **hi2cout** command.

SPI

The SPI bus is a three-wire (plus ground) bus that has two data wires and one clock wire. Data travels in a single direction on each data wire and clock pulses are controlled by the master device. The SPI protocol supports multiple slave devices. Slave devices are addressed in hardware by means of a "chip select" pin, which is controlled by the master device and, in some cases, the slave device may also have an internal address that is fixed at the time of manufacture and/or configured by external address pins. There may be a single or multiple slave devices active at one time, depending on the application. Multiple slave devices may be selected at the same time by daisy-chaining or by internal device addresses; not all devices support these modes of connection.

The data wires are often referred to as MOSI and MISO on master devices and SDI and SDO on slave devices, where MOSI is "Master Out Slave In," MISO is "Master In Slave Out," SDI is "Slave Data In," and SDO is "Slave Data Out." Pull-up resistors are not required for operation, but are sometimes used on the clock and data lines to keep the bus in a known state when it is not being driven.

Electrical connections for the SPI bus with a single slave device are shown in Fig. 3.9.

The PICAXE supports the SPI serial protocol in both software and hardware implementations; the associated commands are:

spiin (shiftin)

spiout (shiftout)

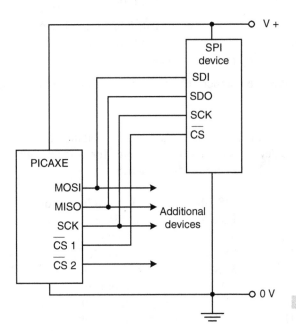

Figure 3.9 Electrical connections for the SPI bus.

hspisetup

hspiin (hshin)

hspiout (hshout)

In addition, it is practicable to write bit-bang routines in PICAXE code and sample routines are given in the PICAXE manual.

The **spiin (shiftin)** command reads from the SPI bus and the **spiout (shiftout)** command writes to the SPI bus.

The **hspisetup** command is used to setup the hardware SPI port on the PICAXE for use by the **hspiin** and **hispiout** commands. The **hspiin** command will read data from the hardware SPI port and the **hspiout** command will write data to the hardware SPI port.

The syntax of the **spiin (shiftin)** command is:

spiin clkpin, datapin, mode, (data {/bits} {, data{/bits} ... })

where

clkpin is a variable or constant in the range 0 to 7 that specifies the pin to use for the clock line

datapin is a variable or constant in the range 0 to 7 that specifies the pin to use for the data line

mode is a variable or constant that specifies the SPI mode.

data is the PICAXE variable(s) that receives the data being read. There may be more than one variable and the number of bytes read from the device will correspond to the number of variables specified in the command.

bits is optional and specifies the number of bits to receive; if not present, the default is 8.

The mode is determined by the slave device and can be obtained from the manufacturer's data sheet for the device being used. The values for *mode* are given in the PICAXE manual and are reproduced here for convenience.

Mode 0	MSBPre_L	MSB first, sample before clock, clock idles low
Mode 1	LSBPre_L	LSB first, sample before clock, clock idles low
Mode 2	MSBPost_L	M SB first, sample after clock, clock idles low
Mode 3	LSBPost_L	M SB first, sample after clock, clock idles low
Mode 4	MSBPre_H	M SB first, sample before clock, clock idles high
Mode 5	LSBPre_H	L SB first, sample before clock, clock idles high
Mode 6	MSBPost_H	MSB first, sample after clock, clock idles high
Mode 7	LSBPost_H	L SB first, sample after clock, clock idles high

The syntax of the **spiout (shiftout)** command is:

spiout clkpin, datapin, mode, (data {/bits} {,data{/bits} ... })

where

clkpin is a variable or constant in the range 0–7 that specifies the pin to use for the clock line

datapin is a variable or constant in the range 0–7 that specifies the pin to use for the data line

mode is a variable or constant in the range 0 to 3 that specifies the SPI mode.

data is the PICAXE variable(s) or constant(s) that is written to the device and there may be more than one.

bits is optional and specifies the number of bits to transmit; if not present, the default is 8.

The mode is determined by the slave device and can be obtained from the manufacturer's data sheet for the device being used. Modes are given in the PICAXE manual and are reproduced here for convenience.

Mode 0	LSBFirst_L	Least significant bit first, clock idles low
Mode 1	MSBFirst_L	Most significant bit first, clock idles low
Mode 2	LSBFirst_H	Least significant bit first, clock idles high
Mode 3	MSBFirst_H	Most significant bit first, clock idles high

The syntax of the **hspisetup** command is:

hspisetup off

hspisetup mode, speed

where

off is a keyword that turns the hardware SPI module off.

mode is a constant or variable that defines the mode to be used.

speed is a constant or variable that defines the speed.

The modes and speeds are given in the PICAXE manual and are reproduced in the following tabulation for convenience.

Mode		Input sampled at	Microcontroller internal settings
spimode00	Mode 0,0	Middle of data time	(CKP=0, CKE=1, SMP=0)
spimode01	Mode 0,1	Middle of data time	(CKP=0, CKE=0, SMP=0)
spimode10	Mode 1,0	Middle of data time	(CKP=1, CKE=1, SMP=0)
spimode11	Mode 1,1	Middle of data time	(CHP=1, CKE=0, SMP=0)
spimode00e	Mode 0,0e	End of data time	(CKP=0, CKE=1, SMP=1)
spimode01e	Mode 0,1e	End of data time	(CKP=0, CKE=0, SMP=1)
spimode10e	Mode 1,0e	End of data time	(CKP=1, CKE=1, SMP=1)
spimode11e	Mode 1,1e	End of data time	(CKP=1, CKE=0, SMP=1)

Speed	Clock divisor	Equivalent speed
spifast	4	1 MHz at 4 MHz
spimedium	16	250 KHz at 4 MHz
spislow	64	63 KHz at 4 MHz

The syntax of the **hspiin (hshin)** command is:

   hspiin (data {, data} ...)

where

   data is the PICAXE variable(s) that receives the data being read. There may be more than one variable and the number of bytes read from the device will correspond to the number of variables specified in the command.

The syntax of the **hspiout (hshout)** command is:

   hspiout (data {, data} ...)

where

   data is the PICAXE variable(s) or constant(s) that is written to the device. There may be more than one variable and the number of bytes written to the device will correspond to the number of variables and/or constants specified in the command.

UNI/O

The UNI/O bus is a single-wire (plus ground) bus that carries data and clock pulses on the same wire. The bus does not require a pull-up resistor for operation; however, a pull-up resistor is normally fitted in order to place the bus in a known state when it is not being driven by a device.

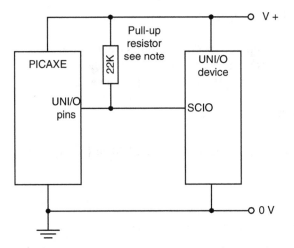

Note. A pull-up resistor is recommended to keep the bus in a known state when it is not being driven by a device. Only one pull-up is required per bus. Typical values are15.5 K to 150 K. 22 K is a nominal value, lower supply voltages require lower values.

Figure 3.10
Electrical connections for the UNI/O bus. The uniin command reads from the UNI/O bus, and the uniout command writes to the UNI/O bus.

The UNI/O bus supports a single master and multiple slaves and the slave address is provided at the time of manufacture. Multiple slaves are supported by using devices that have different addresses.

Electrical connections for the UNI/O bus with a single slave are shown in Fig. 3.10. The PICAXE commands associated with the UNI/O protocol are:

uniin

uniout

The syntax of the **uniin** command is:

uniin pin, devaddr, devcmd, {address, address} (data {,data} …)

where

pin is the PICAXE pin that is used for UNI/O communication.

devaddr is the UNI/O device address that is programmed in the device at the time of manufacture.

devcmd is the command that is being sent to the UNI/O device (see below).

address is the 2-byte address within the UNI/O device and is only required for UNI_READ commands.

data is (are) the PICAXE variable(s) that receive the data being read. Multiple variables may be specified and the device address will be incremented, within page boundaries, for each variable.

There are three device commands that operate with this command, they are:

UNI_READ	Read from the specified address (requires an address)
UNI_CRRD	Read from the current address (uses the next address in the page)
UNI_RDSR	Read status byte (does not require an address)

The syntax of the **uniout** command is:

uniout pin, devaddr, devcmd, {address, address} (data {,data} ...)

where

pin is the PICAXE pin that is used for UNI/O communication.

devaddr is the UNI/O device address that is programmed in the device at the time of manufacture.

devcmd is the command that is being sent to the UNI/O device (see below).

address is the 2-byte address within the UNI/O device and is only required for UNI_WRITE commands.

data is (are) the PICAXE variable(s) that contain the data to be written. Multiple variables may be specified and the device address will be incremented, within page boundaries, for each variable.

There are six device commands that operate with this command; they are:

UNI_WRITE	Write to the specified address (requires an address)
UNI_WREN	Write enable the UNI/O device
UNI_WRDI	Write disable the UNI/O device
UNI_WRSR	Write status
UNI_ERAL	Erase all (write $00 to all locations)
UNI_SETAL	Set all (write $FF to all locations)

Notes:

UNI/O EEPROM devices have a 16-byte page buffer and the memory address will wrap to the beginning of the page when the end of the page has been reached, i.e., the address will increment to $xx0 when address $xxF is reached.

A WREN command must be issued before any writes to the device can take place. The device is automatically write-disabled at power-up and after any of the following commands are successfully executed; WRDI, WRSR, WRITE, ERAL, and SETAL.

The status register may be read to determine if the device is busy and may be written to write-protect specific block of memory. Write-protected blocks cannot be overwritten by any commands until the write-protection is removed.

Examples:

uniout B.4, $A0, UNI_WREN

Write-enable the UNI/O memory chip connected to port B.4, at chip address $A0.

uniout B.4, $A0, UNI_Write, 0, a1, (b0)

Write the byte in variable b0 to memory address 0 in the UNI/O memory chip connected to port B.4, at chip address $A0.

uniin B.4, $A0, UNI_READ, 0, 1, (b1)

Read the byte in the memory address 1 into the variable b1 from the UNI/O memory chip connected to port B.4, at chip address $A0.

1-Wire

The 1-wire bus is a single wire (plus ground) bus that carries data and clock pulses on the same wire, a single pull-up resistor is required for operation. The bus can also be configured to provide power for slave devices making it a truly 1-wire (plus ground) bus. This bus supports a single master and multiple slave devices. Slaves are addressed by means of a unique serial number that is written to the device at the time of manufacture. When there is a single device on the bus, it is not necessary to provide a device address.

Electrical connections for the 1-wire bus and a single-slave device are shown in Fig. 3.11.

The PICAXE implementation of the 1-wire protocol supports a single-slave device; the associated commands are:

readowsn

resetowclk

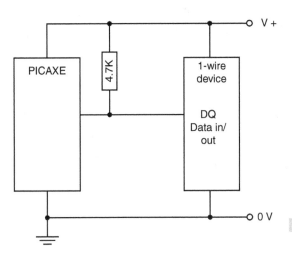

Figure 3.11 Electrical connections for the 1-wire bus.

readowclk

readtemp

readtemp12

The **readowsn** command reads the serial number form any 1-wire device.
The syntax of the **readowsn** command is:

readowsn pin

where

pin is a variable or constant that specifies the input pin to use.

The serial number is read into variables b6 to b13 in the following format:

b6	Family code
b7	'Serial number, least significant byte
b8	'Serial number
b9	'Serial number
b10	'Serial number
b11	'Serial number
b12	'Serial number, most significant byte
b13	Checksum

Example:

readowsn 1

Read the 1-wire serial number from input pin 1 into variables b6 to b13.
The **readowclk** command will read the time in seconds from a DS2415 or DS2417
timer chip, and the **resetowclk** command resets the seconds count to zero. The DS2415
and DS2417 each have a 32-bit counter that is incremented once per second. They are
capable of counting time periods of more than 130 years.
Their syntax is:

readowclk *pin*

resetowclk *pin*

where

pin is a variable or constant in the range 0 to 7 that specifies the input pin to use.

The time (in seconds) is read as a 32-bit binary number that is placed into variables
b10 through b13, where b10 is the most significant and b13 is the least significant
byte.

Example:

resetowclk, 1 'Reset seconds to zero for the DS2415 or DS2417 chip connected to pin1.

readowclk1 'Read seconds into b10, b11, b12, and b13 as a 32-bit unsigned binary number, from the DS2415 or DS2417 chip connected to pin1.

The **readtemp** command reads the temperature in degrees Celsius from a DS18B20 digital temperature sensor into a byte variable as a 7-bit binary integer. The temperature is stored in bits 6 to 0 and bit 7 is the sign indicator. The sign indicator is 0 for positive temperatures and 1 for negative temperatures. Bits 6 to 0 are in the range 0 to 125 for positive temperatures and in the range 1 to 55 for negative temperatures.

The **readtemp12** command reads the temperature in degrees Celsius from a DS18B20 digital temperature sensor into a word variable as a 12-bit signed binary number with eight integer places and four binary places. The sign bit is propagated into the high-order bits of the word and bit 15 is 0 for positive temperatures and 1 for negative temperatures. The temperature range is -55.0000 to $+125.0000$ with a resolution of 0.0625 degrees .

The syntax of the **readtemp** and **readtemp12** commands is:

readtemp pin variable

readtemp12 pin, wordvariable

where

pin is a variable or constant in the range 0 to 7, which specifies the input pin to use.

variable is a byte or word variable that receives the eight-bit result.

wordvariable is a word variable that receives the 12-bit result.

Examples:

readtemp 1, b0

Read the temperature from a DS18B20 digital thermometer connected to pin 1 into b0 as an eight-bit signed binary integer.

readtemp12 2, w0

Read the temperature from a DS18B20 digital thermometer connected to pin 2 into w0 as a 16-bit signed binary number with eight integer places and four binary places.

Keyboard

A PS/2 style keyboard can be connected to a PICAXE chip and the associated commands are:

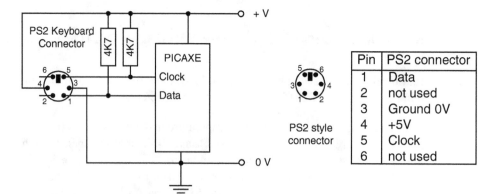

Figure 3.12 Keyboard connections.

keyin

keyled (kbled)

kbin

The **keyin** command reads key codes from a PC keyboard on the PICAXE-18A, -18X, -28A, -28X, and -40X. The **kbin** command reads key codes from a PC keyboard on the X1- and X2-series PICAXE chips. The **keyled** command turns the keyboard indicator lights on and off.

Electrical connections for the keyboard are shown in Fig. 3.12.

The **keyin** command pauses program execution, including interrupt polling, until a keypress occurs and then reads the key code, placing the result into the variable **keyvalue.**

The syntax of the **keyin** command is

keyin

The computer keyboard does not send ASCII characters; instead, it sends numeric codes that represent the keys that have been pressed. It is necessary to convert the key codes into ASCII by means of program code. It is also necessary to track the state of the shift key and convert some keys between upper and lower case, depending on its state.

Example:

keyin

Pause program execution until a key press occurs and place the key code into the system variable **keyvalue**.

The **kbin** command pauses program execution, including interrupt polling, until a key press or timeout occurs. If a key press occurs, the key code is placed into the variable keyvalue; if a timeout occurs, program execution resumes at the label.

The syntax of the **kbin** command is

kbin {[timeout, address,]} variable

where

variable is a byte or word variable that receives the key code

timeout is an optional variable or constant that specifies a timeout period in milliseconds.

label is a label in the program where program execution will resume if a timeout occurs.

Examples:

kbin b0

Pause program execution until a key press occurs and place the key code into the system variable **b0**.

kbin [100, label1], b0

Pause program execution until a key press or timeout occurs. If a key press occurs within the timeout period (100 ms, in this example) the key code is placed in the variable b0. If no key press occurs within the timeout period, then program execution continues at the first command after label1.

The **keyled** command turns a computer keyboard indicator lights on and off, the syntax is:

keyled indicator

where

indicator is a constant or variable that specifies the indicators.

The bit allocations of indicator are:

0 Scroll lock (1 = on, 0 = off)

1 Num lock (1 = on, 0 = off)

2 Caps lock (1 = on, 0 = off)

3 Not used

4 Not used

5 Not used

6 Not used

7 Flashing (1 = no flash, 0 = flash)

Example:

keyled %00000011

Turn flashing off; num lock on and scroll lock on

Infrared

The PICAXE supports the Sony Infrared Command protocol (SIRC); the associated commands are:

infrain

infrain2

infraout

irin

irout

The **infrain** command reads a byte from a 38 KHz infrared receiver attached to an A- or X-series PICAXE chip. The **infrain2** command reads a byte from a 38 KHz infrared receiver attached to an M-, A-, or X-series PICAXE chip. The **infraout** command sends a byte, modulated at 38 KHz, to output port 0 of an M-series chip. The **irin** command reads a byte from a 38 KHz infrared receiver attached to an X1- or X2-series PICAXE chip. The **irout** command sends a byte, modulated at 38 KHz, to a nominated output port of an X1- or X2-series chip.

SIRC uses 12 bit codes that contain seven data bits and five device ID bits. The seven data bits must be in the least significant bits of a byte or word when transmitting and are returned in the least significant seven bits of a byte or word when receiving. Device ID's can be specified with the **infraout** and **irout** commands and must be 1 for communicating with another PICAXE chip.

Some common device IDs are given in the PICAXE manual and are reproduced here for convenience.

Device	ID
PICAXE	1
TV	1
VRT1	2
Text	3
Widescreen	4
MDP/Laserdisk	6
VTR2	7
VTR3	11
Surround sound	12
Audio	16
CD Player	17
Pro-Logic	18
DVD	26

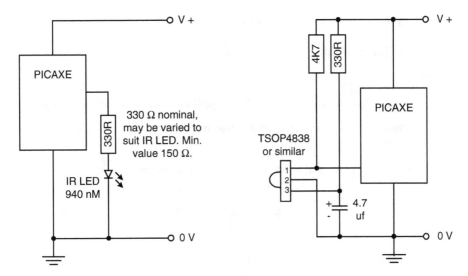

a. Infrared Transmitter   b. Infrared Receiver

Figure 3.13 Infrared interface. (a) Infrared transmitter; (b) intrared receiver.

Electrical connections for the infrared interfaces are shown in Fig. 3.13.

The syntax of the **infrain** command is:

infrain

This command pauses program execution until a byte is received from a 38 KHz infrared receiver and places the received byte into the system variable **infra.**

The syntax of the **infrain2** command is:

infrain2

This command pauses program execution until a byte is received from a 38 KHz infrared receiver and places the received byte into the system variable **infra.**

The syntax of the **infraout** command is:

infraout device, data

where

device is a constant or variable in the range 1 to 31 that specifies the device ID. When transmitting to another PICAXE, the device ID must be 1.

data is a constant or variable in the range 0 to 127 that contains the data to be transmitted.

The syntax of the **irin** command is:

> **irin** {[timeout, address},} pin, variable

where

> *pin* is a variable or constant in the range 0 to 7 that specifies the I/O pin to use.

> *variable* is a byte or word variable that receives the seven-bit data in the least significant bits.

> *timeout* is an optional variable or constant that specifies a timeout period in milliseconds.

> *label* is a label in the program where program execution will resume if a timeout occurs.

The syntax of the irout command is:

> irout pin, device, data

where

> *pin* is a variable or constant in the range 0 to 7 that specifies the I/O pin to use.

> *device* is a constant or variable in the range 1 to 31 that specifies the device ID. When transmitting to another PICAXE, the device ID must be 1. data is a constant or variable in the range 0 to 127 that contains the data to be transmitted.

Pulses

The PICAXE commands that operate with pulses are:

pulsin

count

pulsout

pwm

pwmduty

pwmout

hpwm

hpwmduty

The **pulsin** command will measure the length of a pulse at an input pin. The command pauses program execution until a pulse is detected or a timeout occurs. The pulse can be specified to commence with a high-to-low or low-to-high transition, and the command will timeout if no pulse is detected within a count of 65,535 (655.35 ms at 4 MHz). The command returns the pulse length as the number of 10-μs periods between the leading and trailing edges of the pulse. If the PICAXE is overclocked, the period length must be scaled down proportionally; that is, the period length is 5 μs at a clock frequency of 8 MHz.

The syntax of the **pulsin** command is:

pulsin pin, edge, duration

where

pin is a variable or constant in the range 0 to 7, which specifies the pin.

edge is a variable or constant that specifies the edge transition that begins measurement. 0 (zero) = high-to-low transition, 1 = low-to-high transition.

duration is a word or byte variable that receives the pulse duration in 10-μs increments at 4 MHz.

The **count** command will count the number of positive (low-to-high transitions) pulses at a specific pin during a specific period of time.

The syntax of the **count** command is:

count pin, period, variable

where

pin is a variable or constant that specifies the pin to use

period is a variable or constant that specifies the measurement period in milliseconds at a clock rate of 4 MHz. The period is scaled proportionally with increased clock rate.

variable is a variable that receives the count

Example:

count 2, 1,000, w0

This example will count the number of low-to-high transitions occurring at pin2 during a period of 1,000 ms (at 4 MHz) and place the resulting count into variable w0.

The **pulsout** command generates a single pulse of specific duration at a specified output pin.

The syntax of the **pulsout** command is:

pulsout *pin, time*

where

pin is a variable or constant in the range 0 to 7, which specifies the pin.

time is a variable or constant in the range 0 to 65,535, which specifies the time in 10 μs units at a clock speed of 4 MHz. The time is reduced proportionally at higher clock speeds, i.e., a 10-μs pulse becomes 5 μs at 8 MHz.

The **pwm** command pauses program execution and outputs a pulse-width-modulated signal for a specific number of cycles of approximately 5-ms duration at 4 MHz. The **pwm** command has been replaced by the **pwmout** command and is not discussed further.

The **pwmout** command generates a pulse-width-modulated signal that continues after the command has executed. The signal will only cease when another **pwmout** command is issued with the off keyword or period and duty of 0. The pulse stream will stop during a **nap** or **sleep** command, when the PICAXE is waiting for serial input **(serin, infrain, keyin),** or after an **end** command (express or implied). This command will not operate at the same time as the **servo** command.

The syntax of the **pwmout** command is:

pwmout pin, period, duty cycle

pwmout pin, off

where

pin is a variable or constant that specifies the output pin for the pulse stream.

period is a variable or constant in the range 0 to 255, specifying the period (on time plus off time).

duty cycle is a variable or constant in the range 0 to 1023, specifying the duty cycle (pulse on time).

off is a keyword that turns pwm OFF for the specified pin.

pin may be:

2 for PICAXE-08M and -14M

3 for PICAXE-18M and -18X

C.5 for PICAXE-20X2

(portc) 1 or 2 for PICAXE-28X, -40X, -28X1, and -40X1

C.1 or C.2 for PICAXE-28X2 and -40X2

The **pwmout** command is not available for the PICAXE-20M

The pulse output can be varied by issuing a **pwmduty** command or another **pwmout** command and can be stopped by using the **off** keyword or making the period and duty cycle equal to zero.

The period and duty cycle can be calculated from the "pwmduty" wizard in Programming Editor or from the formulas given in the PICAXE manual. The formulas are reproduced here for convenience.

PWM period = (period + 1) × 4/clock frequency

PWM duty cycle = duty × clock frequency

The 28 and 40X have two pins that can be used for **pwmout** and each may have a different duty cycle, although they must have the same period, because there is only one internal timer that is used for both pins. This internal timer is also used by the **servo** command, which means that the **pwmout** and **servo** commands cannot be used at the same time.

Care must be taken to ensure that the duty cycle (on period) does not exceed the PWM period.

The **pwmduty** command can be used to alter the duty cycle of a previously issued **pwmout** command.

The syntax of the **pwmduty** command is:

pwmduty pin, duty cycle

where

pin is a variable or constant that specifies the output pin for the pulse stream.

duty cycle is a variable or constant in the range 0 to 1023, specifying the duty cycle (pulse on time).

The **hpwm** and **hpwmduty** commands make use of the hardware pwm module that is available in the PICAXE-14M, -28X1, -20X2, -28X2-3V, and -40X2. These commands can operate simultaneously with a **pwmout** command using pin 1, but not when pin 2 is being used.

The syntax of the **hpwm** command is:

hpwm mode, polarity, setting, period, duty

where

mode is a variable or constant that specifies the hardware pwm mode.

polarity is a variable or constant that specifies the active polarity.

setting is a variable or constant that specifies single, half (dead band), or full mode.

period is a variable or constant in the range 0 to 255, specifying the period (on time plus off time).

duty cycle is a variable or constant in the range 0 to 1023, specifying the duty cycle (pulse on time).

The values for mode, polarity, and setting are given in the PICAXE manual and are reproduced here for convenience.

Mode is one of the following:

pwmsingle   0
pwmhalf   1
pwmfull_f   2
pwmfull_r   3

Polarity is one of the following:

pwmHHHH   0
pwmLHLH   1
pwmHLHL   2
pwmLLLL   3

Setting is one of the following:

single mode   %0000 to %1111 to enable/disable DCBA
half mode   dead band delay (0–127)
full mode   not used, use 0 as default value

The syntax of the **hpwmduty** command is:

hpwmduty duty cycle

where

duty cycle is a variable or constant in the range 0 to 1023, specifying the duty cycle (pulse on time).

Servo Motors

The PICAXE commands that control servo motors are **servo** and **servopos.**

The **servo** command allows the PICAXE to drive radio-control style servo motors. The servo output pin is pulsed continuously at about 20-ms intervals, except when the processor is executing a **nap** or **sleep** command or waiting for serial input **(serin, infrain, keyin).** This command will only operate at a clock speed of 4 MHz and will not operate at the same time as the **pwmout** command.

The **servopos** command will adjust the pulse length of a previously issued **servo** command.

Radio-control style servo motors require a pulse of 0.75 to 2.25 ms in duration, which must be repeated continuously at around 50 pulses/second (20 ms) in order for the servo to maintain its position. The electrical connections for radio-control style servos are shown in Fig. 3.14a; the pulses are shown in Fig. 3.14b.

The syntax of the servo command is:

servo port, position

servo {[preload], } pin, position

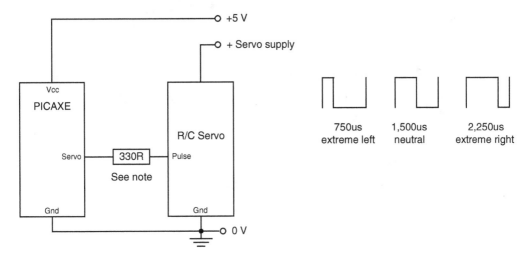

Note. The 330 ohm series resistor is for short-circuit
protection and may be left out in some applications

a. Servo Interface   b. Servo Pulses

Figure 3.14 Servo connections. (a) Servo interface; (b) servo pulses.

where

   port is a variable or constant in the range 0 to 7, which specifies the output pin.

   position is a variable or constant in the range 75 to 255, which specifies the servo
   position.

   preload is an optional timing constant for X2 chips only and is used to vary the pulse
   rate.

The **servo** command will only function at clock speeds of 8 or 32 MHz for X2 parts, 4
or 16 MHz for X1 parts, and 4 MHz for other parts.
   The syntax of the **servopos** command is:

servopos port, position

where

   port is a variable or constant in the range 0 to 7, which specifies the output pin.

   position is a variable or constant in the range 75 to 255, which specifies the servo
   position.

A value of 150 for *position* causes the servo to move to center position, a value of
75 causes the servo to move to maximum displacement on one side, and 225 causes
maximum displacement to the other side. These values can vary marginally from servo
to servo and are for servos that require a 1.5-ms neutral pulse.

Examples:

servo 2, 150 'Position the servo on port 2 to center or neutral position

servo 2, 75 'Position the servo on port 2 to maximum displacement on one side

servo 2, 225 'Position the servo on port 2 to maximum displacement on the other side

servopos 2, 110 'Reposition the servo on port 2 to approximately one-half displacement

4

PROGRAMMING

What Is a Program?

A program is a set of instructions that tells a microcontroller what to do. Programs can be very simple, such as turning LED on and off, or complex, such as controlling a digital clock/calendar.

For a program to run on a microcontroller, it must be in the form of binary machine code that the microcontroller is designed to understand. Such binary machine code is not particularly suitable for most human beings who prefer a language that they can understand more easily. For this purpose, a number of different programming languages have been developed. These programming languages allow the programmer to write code in a language that is more easily understood by humans, and computer software called a compiler is then used to translate the program in to a binary form that the microcontroller understands. (Compilers have many other features such as text editing, syntax checking, and debugging.)

There are several compilers available for the PICAXE; some are listed in Table 4.1. These compilers provide the programmer with a choice of writing a program in BASIC code or by drawing the program as a flowchart. All these software packages are available on the programming editor CD-ROM (part BAS805) and can be downloaded from the Revolution Education website.

Writing BASIC Code

PICAXE program code is written in a language called BASIC (which stands for Beginner's All-Purpose Symbolic Instruction Code). BASIC is a very simple language to learn and apply. It is typed using commands that make use of data, constants, operators, expressions, and keywords. Compiler directives and comments may also be included with the program.

TABLE 4.1 PICAXE COMPILERS

SOFTWARE PACKAGE (COMPILER)	PROGRAMMING OPTIONS	LICENSE	SUPPORTED PLATFORMS
Programming editor	BASIC code Flowcharting	Free	Windows
Logicator for PIC	Flowcharting	Site license required Free for single user	Windows
AXEpad	BASIC code	Free	Windows Macintosh Linux

Compiler Directives

Compiler directives are used to provide information to the compiler about the program that is being compiled. Directives do not become part of the program, although they may affect its content, and do not take up any memory space in the PICAXE chip. Note also that the **EEPROM, data,** and **TABLE** commands operate in a similar way to directives in that they do not take up space in PICAXE program memory.

Directives commence with the pound or hash symbol (#), which is then followed by one or more keywords. Directives are listed in the PICAXE manual; some are reproduced in Table 4.2 for convenience.

Data

Data is used by program commands during the operation of a program and consists of variables, constants, or expressions. Data is generally stored in an area of memory that is reserved for that purpose, although it may be stored in program memory in some circumstances. Variables are items that may be changed in value by commands. Constants are items that are used by commands, but cannot be changed. Expressions are combinations of variables, constants, and operators that return a value.

Variables may be words, bytes, or bits. Words consist of 16 bits, bytes consist of 8 bits, and bits consist of single binary digits. Word variables are addressed by the letter "W," which may be upper or lower case, followed by a number specifying the word. Byte variables are addressed by the letter "B," which may be upper case or lower case, followed by a number specifying the byte. Bit variables are addressed by the letters "bit," which may be upper case or lower case, followed by a number specifying the bit. Variables may also be addressed by means of pointers and may be given alternate names by means of the **symbol** command. Note also that words, bytes, and bits occupy the same memory space. More information can be found in Chapter 2, PICAXE Architecture.

Examples:

W0	Word 0 (zero)
w1	Word 1
b0	Byte 0 (zero)
B1	Byte 1
Bit0	Bit 0 (zero)
bit7	Bit 7
@ptr	A byte in the scratchpad pointed to by the scratchpad pointer.
@bptr	A byte in the byte-scratchpad pointed to by the byte-scratchpad pointer.

Constants are used for data that does not change. They can be numeric, character, or string and can be used in nearly any place that a variable can be used, except that a constant cannot be placed to the left of the assignment operator (= sign). Numeric constants can be written as a decimal number, hexadecimal number, or binary number.

TABLE 4.2   COMPILER DIRECTIVES

DIRECTIVE	OPERATION	EXAMPLE
#picaxe xxx	Set the compiler mode for a specific PICAXE chip where *xxx* identifies the chip type.	#picaxe18x #picaxe28x2
#com device	Specify the communications port to be used for program download where *device* identifies the port.	#com COM2 #com /dev/ttyS0 #com /dev/tty.usbserial-xxxx
#slot nn	Set the PICAXE (or EEPROM) slot number to receive the downloaded program where *nn* is the slot number.	#slot 1 # slot 4
#revision nnn	Set the revision number for the program where *nnn* is the revision number of the program	#revision 7
#no_data	Do not download data to the EEPROM.	#no_data
#no_table	Do not download data to the TABLE or EEPROM.	#no_table
#freq mn	Set the clock frequency (X parts only) where *n* is the frequency.	#freq m4 #freq m16
#define label	Define a compiler label, where *label* is the label to be defined.	#define anylabel
#undefine label	Remove the definition of a compiler label, where *label* is the label to be undefined.	#undefine anylabel
#ifdef label {#else} #endif	Conditionally compile code depending on the presence of a label, where *label* is the label. The #else directive is optional.	#ifdef anylabel Place code here to be compiled if "anylabel" is defined. #else Place code here to be compiled if "anylabel" is not defined. #endif
#ifndef label {#else} #endif	Conditionally compile code depending on the absence of a label, where *label* is the label. The #else directive is optional.	#ifndef anylabel Place code here to be compiled if "anylabel" is not defined. #else Place code here to be compiled if "anylabel" is defined. #endif
#error text	Force a compiler error, where *text* is the error message to be displayed.	#error Error message
#rem / #endrem	Comment out multiple lines of text	#rem Code placed here will be treated as a comment and will not be compiled. #endrem

(continued)

TABLE 4.2   COMPILER DIRECTIVES (Continued)

DIRECTIVE	OPERATION	EXAMPLE
#include	This directive is not currently implemented. Include code from an external file.	#include "c:/codelib.bas"
#slot1file filename #slot2file filename	Include the file name of other slots for simulations.	#slot1file "c:/slot1code.bas"
#gosubs nnn	Set the gosubs mode for a program, where *nnn* is the mode.	#gosubs 16 #gosubs 256
#sim project	Display a project on the screen while simulating a program.	#sim axe102 #sim axe092
#simspeed nnn	Set the speed of program simulation, where *nnn* is the delay between iterations in milliseconds.	#simspeed 250
#terminal off #terminal nnnn	Configure the programming editor terminal to open at a specific baud rate, where *nnnn* is the baud rate.	#terminal 2400

Decimal constants contain the digits 0 to 9, hexadecimal constants are preceded by a dollar sign ($) and contain the digits 0 to 9 and A to F, and binary constants are preceded by a percent sign (%) and contain the digits 0 or 1. String constants contain alphanumeric information and are enclosed between double quote marks ("). String constants are stored in program memory when they are part of a command, e.g., **serout** 1, N2400, ("ABCDEF"). EEPROM and the TABLE can also be used for storing string constants.

Examples:

145	The decimal constant one hundred and forty-five
$2A	The hexadecimal constant 2A
%00101010	The binary constant 00101010
"A"	The ASCII character A
"ABCDEF"	The ASCII string ABCDEF
"235"	The ASCII string 235
"$1B"	The ASCII string $1B
"%10"	The ASCII string %10

Any character, except the double quote mark ("), can be used in a character or string constant. The double quote mark and control characters, such as carriage return and line feed, can be referred to by their hexadecimal values.

Expressions are combinations of variables, constants, and operators that evaluate to a value. They can be used in the assignment (**let**) command and some other commands. An expression takes the following form:

variable operator variable/constant {operator variable/constant ... }

where

variable is one of the PICAXE word, byte, bit, special variables, or pointers.

operator is one of the arithmetic or Boolean operators.

constant is a number or character.

The arithmetic and Boolean operators are listed in the PICAXE manual and are discussed in Chapter 5, Arithmetic and Data Conversion.

Expressions are evaluated from left to right. The expression $1 + 2 * 3$ will be evaluated to 9 ($1 + 2 = 3 * 3 = 9$), not 7, which may be expected from the BODMAS rules of mathematics that apply multiplication before addition. BODMAS is an acronym that is used to specify the order of evaluation of mathematical expressions and sometimes it is expressed with different letters. It stands for Brackets, Other (exponentiation and trigonometric functions), Division, Multiplication, Addition, Subtraction.

Meaningful Names

The default names for variables do not give much indication as to what data the variables contain, other than that they are a word, byte, or bit, and this can make it difficult to read program code. To overcome this, the **symbol** command can be used to give a meaningful name to variables, constants, and other items. Symbol commands are usually placed near the beginning of a program and once a symbol has been defined it can be used anywhere in a program where its context is valid. Symbol commands are used by the compiler and do not take up any space in PICAXE memory.

The syntax of the **symbol** command is:

symbol symbolname = value {operator constant}

where

symbolname is the name to be given to the symbol.

value is any constant or variable that is valid in the context of the symbol.

operator is an optional arithmetic or Boolean operator. Only one operator is allowed.

constant is an optional number or predefined symbol.

The following rules must be followed when naming a symbol:

Symbol names must begin with a letter or an underscore (_).

Symbol names may contain letters, numbers, or an underscore.

Symbol names must not consist of a reserved word.

Symbol names are not case sensitive, and "dataout" and "DATAOUT" are the same.

Symbols must be defined before they can be used.

Examples:

symbol outbyte = b4	"outbyte" can be used in place of b4
symbol wordcount = w2	"wordcount" can be used in place of w2
symbol signbit = bit7	"signbit" can be used in place of bit7
symbol dataout = 0	"dataout" can be used in place of 0
symbol switch = pin3	"switch" can be used in place of pin3
symbol symbol_2 = %11010101 & $0F	"symbol_2" = %00000101

Commands

Commands are the instructions that cause the PICAXE to carry out operations. Commands may have a single keyword, such as, **high, pause,** and **let,** or they may have pairs or sets of keywords such as **for ... next, gosub ... return, do ... loop, if ... then ... else ... endif,** and **select ... case ... {case ... } else ... endselect.** Each command has its own rules (syntax) and may require parameters consisting of variables, constants, or expressions to be specified by the programmer. The programming editor command syntax is given in the PICAXE manual, which can be found in the "help" menus of programming editor and AXEpad.

Labels

Labels are names that are used as reference points within a program. They are defined by the programmer at the time a program is being written. The use of structured commands such as **if ... then ... else ... endif,** and **do ... loop** reduces the need for labels; however, there are still places where labels must be used, such as, defining the entry point for subroutines. Labels do not take up any space in PICAXE memory.

The following rules must be followed when naming a label:

Label names must begin with a letter.

Label names may contain letters, numbers , or an underscore (_).

Label names must end with a colon (:).

Label names must not consist of a reserved word.

Label names are not case sensitive, and "label1:" and "LABEL1:" are the same.

Label names must be unique.

[When label names are used in commands, the colon (:) is not used.]

Labels are usually written beginning in column 1 of the editor workspace and may be written on a line by themselves or on the same line as a command.
Examples:

label1: command	This label is on the same line as a command
label2:	This label is on a line by itself
goto label1	Resume program execution at the first command after label1. Note that the colon character is used only when defining labels and is not used when referring to them.

Comments

Comments are text that may be inserted into a program by the programmer and are usually used for documentation. The single quote mark ('), semicolon (;), or the word "REM," which may be in upper or lower case, are used to begin a comment and all text on a line after the quote mark, semicolon, or "REM" is ignored by the compiler. Comments may be on a line by themselves or at the end of a line and do not take up any space in PICAXE memory. It is not necessary to use a comment to insert a blank line in a program because blank lines are ignored by the compiler.

Comments may also be enclosed between the compiler directives **#rem** and **#endrem.**
Examples:

'This comment begins with a single quote mark

label1:; this comment begins with a semicolon and is on the same line as a label

pause 200 rem This is a comment that starts with "rem"

'pause 200 This is a comment that uses a line of its own, "pause 200," although a valid command, is part of the comment and is ignored by the compiler.

#rem	Begin a comment
high 0	"high 0" is part of the comment and is ignored by the compiler
	More lines of code/text can be placed here and will be part of the comment
#endrem	End a comment

Reserved Words

Reserved words are the names of labels, commands, system variables, and other items that are reserved by the compiler for its own use. Reserved words must only be used in the context that they are intended for and cannot be used in **symbol** commands or as labels. The words that are reserved are listed in the PICAXE manual and sometimes change when new versions of compilers are released. If a program is causing a compile error for no apparent reason on a line containing a symbol command or a label, look for a reserved word being misused.

Assigning Values to Variables

The **let, inc,** and **dec** commands are most often used to assign values to variables, although there are many other commands that will place values in variables including **serin, readi2c,** and **readowsn.**

The syntax of the **let** command is:

{**let**} variable = {-} value

where

let is optional and can be left out.

variable is one of the PICAXE word, byte, bit, or special variables (including pointers).

value is a constant, variable, or expression that returns a value.

- is an optional minus sign that will cause the value to be negative.

The syntax of the **inc** and **dec** commands are:

inc variable

dec variable

where

variable is one of the PICAXE variables.

Examples:

let b0 = 1	'b0 is assigned the value 1
b0 = −1	'b0 is assigned the value minus 1
w0 = w1	'The value in w1 is placed in w0
b0 = b0 + 1	'The value in b0 is added to the constant 1 and the 'result is placed in b0
b4 = b5 + b6	'The value in b5 is added to the value in b6 and the 'result is placed in b4
w0 = b4	'The value in b4 is placed in w0 and unused high-order bits are cleared to 0
b4 = b2 * b1	'The value in b2 is multiplied by the value in b1 and the low byte of the result is placed in b4
b0 = b0 & $0F	'The value in b0 is logically ANDed with the hexadecimal constant 0F and the result is placed in b0
inc b0	'The value in b0 is incremented by 1
dec w2	'The value in w2 is decremented by 1

Advanced Programming

PROGRAM FLOW

Program execution starts with the first command in a program and then proceeds to the following command unless a command, such as **goto,** changes the order of flow. This process will continue until an **end** command is reached or the last command in the program is executed.

Commands that directly change the order of program execution are:

goto

gosub

on ... goto

on ... gosub

branch

for ... next

do ... loop

exit

return

Commands that can change the order of program execution indirectly are:

if ... then

case ... endcase

irin

kbin

serin

setint

setintflags

The commands that change the order of program execution indirectly do so by means of commands that directly change the order of execution being placed within them (**if ... then, case ... endcase**), having timeout facilities that continue program execution at a specified label (**kbin, serin**), or by setting an interrupt condition that causes execution to move to a specific label when an interrupt occurs (**setint, setintflags**).

CODE STRUCTURES

When writing programs, we tend to write code in one of three different structures: sequence, selection, and iteration (looping).

Sequences are commands that are written one after the other as in:

high 2

pause 200

low 2

pause 200

SELECTION

Selection commands are commands that change the order of program execution, depending on a condition.

The PICAXE commands that can perform selection are:

select ... endselect

if ... then ... endif

branch

on ... goto

on ... gosub

The **select ... endselect, if ... then ... endif, branch, on ... goto**, and **on ... gosub** commands can cause program execution to take one of several paths or continue at one of several different places.

The syntax of the **select ... endselect** command is:

select variable

case expression {**code**}

{**case** expression {**code**} ... }

{**else** code}

endselect

where

variable is one of the PICAXE variables.

expression is an expression that is valid in context.

code is valid PICAXE code.

There may be more than one **case** keyword, the **else** keyword is optional, and the code is optional for a particular case. The optional features of this command allow it to take many different forms and the command is frequently written using several lines.

Example:

b1 = 10

select b0

    case 0 serout 1, T2400, ("b0 = 0")

    case 1, 2, 3 serout 1, T2400, ("b0 = 1, 2, or 3")

    case 4 to 9 serout 1, T2400, ("b0 = 4 to 9")

    case b1 serout 1, T2400, ("b0 = b1 which is 10")

    case > 50 serout 1, T2400, ("b0 is greater than 50")

else serout 1, T2400, ("None of the above cases was true, won't happen for this example")

endselect

This example will send different values to a serial port depending on the value of variable b1.

The syntax of the **if ... then ... endif** command is:

if condition **then code**

{**elseif** condition **then code**}

{**else code**}

{**endif**}

where

    condition is an expression that evaluates to a Boolean (true or false) value.

    code is valid PICAXE code.

If the condition is evaluated as true, program execution continues with the first command after the **then** keyword, otherwise program execution continues with the command following the **else** keyword if it is present, or the **endif** keyword if an **else** keyword was not present. The **elseif, else,** and **endif** keywords are optional, although the **endif** keyword must be present if there is an **elseif** or **else** keyword or if there is **code** other than **goto.** The optional features of this command allow it to take many different forms and the command is frequently written using several lines.

Examples:

if b0 = 10 then label1

if bit4 is 0 then label3

if bit1 = 0 and bit5 = 1 then label4

if pin3 = 0 then label5

if w0 = 10000 then high 2 endif

if pin2 = 0 then high 1 else low 1 endif

The syntax of the **branch** command is

branch offset, (label0, label1, label2, . . . labeln)

where

offset is variable or constant.

label0 is the label after which execution commences if variable = 0.

label1 is the label after which execution commences if variable = 1.

label2 is the label after which execution commences if variable = 2.

labeln is the label after which execution commences if variable = n.

Each label corresponds to one of the possible values of the offset. If the value of the offset is greater than the number of labels specified, then program execution continues with the command following the **branch** command.

ITERATION

Iteration is the process of performing a command, or group of commands, multiple times and is often called looping. The PICAXE commands that perform iteration are **do . . . loop** and **for . . . next.** In addition, loops can be constructed from other commands that change the order of program execution, such as **if . . . then . . . else** and **goto.**

The **do . . . loop** command allows a command, or group of commands, to be repeated, while a specific condition exists or until a specific condition occurs. The **for . . . next** command allows a command, or group of commands, to be repeated a fixed number of times.

The syntax of the **do . . . loop** command is:

do {**until/while** condition}

code

loop {**until/while** condition}

where

condition is an expression that evaluates to a Boolean (true or false) value.

code is valid PICAXE code.

The condition is optional and if not specified in either the **do** command or **loop** command, then the loop will repeat continuously.

Examples:

```
do                   'No exit condition is specified, so the loop will repeat forever
                     'Code goes here
loop
do while b0 <> 50    'Specify an exit condition on the do line
     inc b0
loop

do
inc b0
loop until b0 = 50   'Specify an exit condition on the loop line
```

The **exit** command can be used within a **do ... loop** command to exit the loop immediately.

The syntax of the **for ... next** command is:

for variable = startvalue **to** endvalue {**step** {-}increment}

code

next {variable}

where

variable is one of the PICAXE word or byte variables.

startvalue is a variable or constant value to which *variable* will be initialized.

endvalue is a variable or constant value that controls when the command will end.

step is an optional clause that specifies an increment.

increment is a variable or constant value by which *variable* will be incremented each time program execution reaches the **next** command. *Increment* is optional and may be positive or negative; if not present, it defaults to 1.

code is valid PICAXE code.

The **for ... next** command operates by initializing *variable* to *startvalue*. The commands in the loop are then executed in logical order beginning with the first command after the **for** command. When the **next** command is reached, the end condition is checked. If the end condition is reached, program flow moves to the command after the **next** command, otherwise program flow moves to the first command after the **for** command. The end condition will occur when *variable* would be out of range if another iteration of the loop occurred.

Notes:

The *variable* is optional in the **next** command and if not present, the **next** command is associated with the previous **for** command. If *variable* is present in the **next** command, then the **next** command is associated with the **for** command that uses the same variable.

for ... next loops can be nested within each other up to eight levels.

The values of *variable, startvalue, endvalue,* and *increment* should not be changed by any commands during the **for ... next** loop. The value of *variable* can be read, and often is, during the execution of the loop.

The **exit** command can be used within a **for ... next loop** to exit the loop immediately.

Examples:

for b0 = 1 to 5	'Set up a loop that executes 5 times
if b0 = 4 then exit	'Exit on the 4th iteration
next	
for b0 = 1 to 10	'Loops 10 times
for b1 = 200 to 100 step -10	'Loops 20 times for each iteration of b0 loop
command(s)	'These commands will execute 200 times
next b1	
next b0	

Loops can also be constructed using **goto** and **if ... then** commands.

Examples:

label1: goto label1	'Endless loop that does nothing other than demonstrate looping
b0 = 1	'Initialize a counter
label1: inc b0	'Increment the counter
if b0 < = 4 then label1	'Test for exit

POINTERS

Pointers are variables that contain the address of another variable; they are discussed further in Chapter 2.

SUBROUTINES

Subroutines are blocks of code that can be reused in different places in a program. They can reduce the size of a program and can also be used to make program code more readable. The PICAXE command that operates with subroutines is **gosub ... return.** The number of **gosub** commands in a program is limited to 15 or 255, depending on the PICAXE chip and the programming editor options. Subroutines can be nested up to four or eight levels depending on the PICAXE chip.

The **gosub** command operates by causing the first command after the *label* to be the next command that is executed. It also remembers its place in the program and when a **return** command is executed, program execution will return to the first command after the most recently executed **gosub** command.

The syntax of the **gosub** command is:

gosub *label*

return

Examples:

command1

gosub sub1

command2

gosub sub1

command3

gosub sub2

command4

gosub sub1

command5

end

sub1: *command6*

return

sub2: *command7*

return

Note that *command1, command2, command3, command4, command5, command6,* and *command7* may consist of more than one command.

In this example, the order of program execution is:

command1 (because it is the first command in the program)

gosub *sub1* (because it is the next command)

command6 (because it is the first command after the label *sub1*)

return (because it is the next command)

command2 (because it is the first command after the most recent **gosub**)

gosub *sub1* (because it is the next command)

command6 (because it is the first command after the label *sub1*)

return (because it is the next command)

command3 (because it is the first command after the most recent **gosub**)

gosub *sub2* (because it is the next command)

command7 (because it is the first command after the label *sub2*)

return (because it is the next command)

command4 (because it is the first command after the most recent **gosub**)

gosub *sub1* (because it is the next command)

command6 (because it is the first command after the label *sub1*)

return (because it is the next command)

command 5 (because it is the first command after the most recent **gosub**)

end (because it is the next command)

```
gosub sub1
end
sub1:
gosub sub2
command(s)1
return
sub2: command(s)2
return
```

In this example, *sub2* is nested within *sub1* and the order of execution is:

gosub sub1 (because it is the first command in the program)

gosub sub2 (because it is the first command after the label sub1)

command(s)2 (because it is the first command after the label sub2)

return (because it is the next command)

command(s)1 (because it is the first command after the most recent **gosub**)

return (because it is the next command)

end (because it is the first command after the most recent **gosub**)

PAUSING AND STOPPING PROGRAM EXECUTION

The commands that directly pause or stop program execution are:

end

nap

sleep

pause

pauseus

stop

wait

doze

hibernate

reset

There are also some commands that indirectly pause program execution, such as **serin, infrain,** and **keyin.**

The **end** command stops program execution and places the PICAXE into a low-power mode until a hardware reset occurs. The **end** command is optional and, if present, is often the last command in the program, if not present an **end** command is assumed after the last command in the program.

The **nap** and **sleep** commands stop program execution and place the PICAXE into a low-power mode for a period of time. When the time period expires, power is restored to normal and program execution continues from the point where it left off. The time period is independent of the PICAXE clock frequency and can vary with temperature and different PICAXE chips by as much as 100%. These commands are often used to save power when the PICAXE is powered from a battery; however, they should not be used when accurate timing is required. Most of the internal components of the PICAXE are disabled during the **nap** and **sleep** commands and the **pwmout** and **servo** commands do not function. The difference between the **nap** and **sleep** commands is the length of time delay. **nap** can delay up to 2.3 seconds; **sleep** can delay for over 40 hours in 2.3-second increments.

The syntax of the **nap** command is:

nap *time*

where

time is a variable or constant in the range 0 to 7.

A value of

0 gives a delay of 18 ms.

1 gives a delay of 32 ms.

2 gives a delay of 72 ms.

3 gives a delay of 144 ms.

4 gives a delay of 288 ms.

5 gives a delay of 576 ms.

6 gives a delay of 1152 ms.

7 gives a delay of 2304 ms.

Example:

nap 3

This command will cause the PICAXE to suspend program execution and enter low-power mode for approximately 144 ms. At the end of this period, the chip will revert to normal power mode, and program execution will continue where it left off.

The syntax of the **sleep** command is

sleep *time*

where

time is a variable or constant in the range 0 to 65,535, which specifies the time period in multiples of 2.3 seconds.

Examples:

b0 = 10

sleep 1 'Sleep for approx. 2.3 s

sleep b0 'Sleep for approx. 23 s

sleep 7826 'Sleep for approx. 5 h

The **pause** command will suspend program execution for a period of time that is specified in milliseconds, the **pauseus** command is available on X1- and X2-series chips and will suspend program execution for a period of time that is specified in increments of 10 μs. The time period is as accurate as the PICAXE clock and is proportional to the clock rate.

The syntax of the **pause** and **pauseus** command is:

pause *timems*

pauseus *timeus*

where

timems is a variable or constant in the range 0 to 65,535 that specifies the time delay in ms at a clock rate of 4 MHz. The time is reduced proportionally with higher clock rates.

timeus is a variable or constant in the range 0 to 65,535 that specifies the time delay in increments of 10 μs at a clock rate of 4 MHz on X1 parts and 8 MHz on X2 parts. The time is reduced proportionally with higher clock rates.

Examples:

setfreq m4	'Set the clock to 4 MHz
pause 200	'Pause for 200 ms
pauseus 10	'Pause for 100 μs
b0 = 500	
pause b0	'Pause for 500 ms.
pauseus b0	'Pause for 5000 μs
setfreq m8	'Set the clock to 8 MHz
pause 100	'Pause for 50 ms (delay is halved at 8 MHz)
pauseus 1	'Pause for 5 μs (time periods this short may be swamped by the command execution time)

The **stop** command stops program execution until a hardware reset or a program download occurs. It does not place the PICAXE into low-power mode, and **pwmout** and **servo** commands will continue to operate.

The syntax of the **stop** command is:

stop

The **wait** command will suspend program execution for a period of time that is specified in seconds It is functionally equivalent to the **pause** command, except that the time must be a constant and is specified in seconds rather than milliseconds. The time period is as accurate as the PICAXE clock and is proportional to the clock rate.

The syntax of the **wait** command is:

wait *time*

where

time is a constant in the range 0 to 65, which specifies the time delay in seconds.

Examples:

setfreq m4 'Set the clock to 4 MHz

wait 2 'Wait for 2 s

setfreq m8 'Set the clock to 8 MHz

wait 2 'Wait for 1 s (delay is halved at 8 MHz)

The **doze** command stops program execution for a period of time or until an interrupt occurs on the X2-series chips. The time period is independent of the PICAXE clock frequency and can vary with temperature and different PICAXE chips by as much as 100%; the command should not be used when accurate timing is required. The internal components of the PICAXE continue to function during the **doze** command and the **pwmout** and **servo** commands will continue to operate.

The syntax of the **doze** command is:

doze time

where

time is a variable or constant that specifies the period of time in 2.1-ms increments.

The **hibernate** command stops program execution and places the X1-series chips into a low-power mode, which will continue until a reset, interrupt, or a manual wakeup occurs. Manual wakeup occurs when hibernate mode has been enabled and the ACD0 pin is brought low by either a capacitor discharge or manual switch.

The syntax of the **hibernate** command is:

hibernate mode

where

mode is 0 to disable manual wakeup or nonzero to enable manual wakeup

The **reset** command will perform a hardware reset on the X1- and X2-series chips. The syntax of the **reset** command is:

reset

POWER SAVING

Many of the commands that pause program execution also place the PICAXE into a low-power mode and this feature can be used in certain circumstances to save power. For example, a data-logging situation in an orchard may take a temperature reading every 3 min and the PICAXE can be placed into a low-power mode between readings to save power.

INTERRUPTS

Interrupts must be enabled with either the **setint** or **setintflags** commands prior to use. The program must contain an interrupt service routine that is written as a subroutine at the end of the program commencing with the label "interrupt:" and ending with a **return** command.

When an interrupt occurs, interrupts are disabled and program execution is transferred to the first command after the label "interrupt:". The effect is equivalent to executing "setint(flags) off" and "gosub interrupt" commands. To return to the main program, a **return** command must be executed within the interrupt subroutine. Interrupts are automatically disabled when an interrupt occurs and must be re-enabled if another interrupt is to be processed.

A test is made for interrupt conditions between each command, between each note of a tune, and during **pause** commands. Interrupt conditions are not tested during serial input commands such as **serin, infrain,** and **keyin.** A small amount of time is taken to test for the interrupt condition and this will have a minor effect on programs and the time delay of a **pause** command is slightly extended.

It is generally a good idea to keep the interrupt routine processing to a minimum to avoid affecting the main program. It is also important to ensure that the interrupt condition has passed before returning from the interrupt routine to ensure that another interrupt does not occur for the same condition.

The **setint** command enables interrupts that depend on the state of pins. The **serintflags** command enables interrupts that depend on the state of the flags in the system variable **flags.**

The syntax of the **setint** command is:

SETINT off

SETINT {**not**} condition, mask {, port}

where

off is a keyword that turns interrupts off.

not is an optional keyword on X1 and X2 parts that inverts the state of the *condition.*

condition is a variable or constant that specifies the state of the pins that define the interrupt. If a bit is 0 then an interrupt will occur if the corresponding pin is at logic low level and the corresponding mask bit is a 1. If a bit is 1, then an interrupt will occur if the pin is at logic high level and the corresponding mask bit is a 1.

mask is a variable or constant that defines the pins that are examined for the interrupt condition. If a mask bit is a 0, then the corresponding pin will not be examined, if a mask bit is a 1, then the corresponding pin will be examined.

port is an optional variable or constant on X2 parts that defines the port that will be examined for an interrupt condition, e.g., A, B, C, D.

The syntax of the **setintflags** command is:

setintflags off

setintflags {not} flags, mask

where

not is an optional keyword that inverts the state of the flags.

flags is a variable or constant that specifies the flags that will generate an interrupt.

mask is a variable or constant that defines the flags that are examined for the interrupt condition. If a mask bit is a 1, then the corresponding flag will be examined.

The flags are the components of the system variable "flags" and are discussed further in Chapter 2, PICAXE Architecture.
    Example:

```
setint %00010000, %00010000       'Enable interrupts on pin4 high
do
    'put program code here
loop
interrupt:                        'Interrupt routine
    'put interrupt routine code here
do while pin4 = 1 loop            'Optionally wait for interrupt condition to pass
pulsout 0, 1                      'Optionally send a pulse to reset the interrupt
                                   condition
setint %0010000, %0010000         'Optionally enable interrupts again
return                            'Return from interrupt routine
```

PROGRAM TESTING AND DEBUGGING

Let's face it, no programmer is perfect and even the simplest of programs may not perform as intended at the first test. Faults in programs are called "bugs." The process of isolating and correcting faults is called "debugging."
    Programming editor, AXEpad, and Logicator for PIC all have inbuilt debugging facilities to assist programmers to identify bugs in their programs; Table 4.3 shows the capabilities of each.
    The **debug** command transfers data from the PICAXE to the software via the programming link, which takes a finite amount of time so the program will run more slowly

TABLE 4.3 DEBUG FACILITIES		
SOFTWARE	**DEBUG FACILITIES**	
Programming editor	Debug command	Simulate menu option
AXEpad	Debug command	Debug menu option
Logicator for PIC	Debug command	Simulate menu option

when the **debug** command is used. The big advantage of the **debug** command is that the value of variables is displayed on the screen in real time.

It may not always be convenient or practicable to use the **debug** command to locate faults in a program and other methods of debugging programs can be used, such as turning LEDs on or off to indicate that the program has taken a particular path, that a variable contains a particular value, or an input port is in a particular state. The **serout** command can also be used to display messages on an LCD.

The syntax of the debug command is:

debug {variable}

where

variable is optional and has no effect on the command.

The **debug** command can be placed anywhere in a program and it will send the status of variables via the programming link to the PC. The ability to view the contents of the variables as your program executes can help to reveal the point(s) at which your program is not performing as intended (a.k.a. doing the wrong thing). The PICAXE system being debugged must be connected via the programming port to a PC running the programming software. A modified serial programming circuit is recommended if the program contains any **readadc10** commands. The debug command slows down the execution of a program and should be removed when the program is working properly.

If the debug command does not give sufficient information to be able to isolate a problem then the simulate feature in programming editor or Logicator for PIC can be used.

Program simulation is a menu option in programming editor and Logicator for PIC that allows program code to be visualized on the screen. Simulation allows a program to be executed a step at a time and the values of variables and the states of input/output ports can be displayed. Commands can be executed one at a time by stepping, or breakpoints can be set to allow the program to proceed to a particular point before simulation starts.

The simulate feature of programming editor includes the following features:

Run

Stop

Single step

Break

Reset

Variable breakpoint

Line breakpoint

Highlight the line of code that is executing

Display variables in binary hex and decimal.

Display memory in hex and decimal (Data, SFR, Scratchpad, Table)

Display output pins states, set input pin states

Display the serial output buffer

The simulate feature of Logicator for PIC includes the following features:

Run

Reset

Stop

PROGRAMMING WITH FLOWCHARTS

Programming editor and Logicator for PIC both have facilities to write programs flowcharts rather than by writing code and this can be a very effective way to introduce oneself, or students, to the art of programming.

5

PICAXE ARITHMETIC AND

DATA CONVERSION

CONTENTS AT A GLANCE

Number Systems

Internally, microcontrollers work with binary values. Human beings usually do not find the binary system easy to work with and tend to use other number systems, such as octal, hexadecimal, decimal, or binary-coded decimal (BCD).

BINARY

The binary system has two distinct digits: 0 and 1 that are called bits (binary digits). Individual digits are written next to each other to make up a number. If the number has a fractional part, the integer portion and the fractional portion are usually not separated. Instead the number of fractional digits is specified. Each binary digit is related to its neighbor by a factor of 2.

The binary system is easy to implement in electronic circuits and most computers use this system or a system that is derived from it, such as BCD. Bits (binary digits) do not normally exist by themselves, instead they are usually grouped together into bytes (eight bits) or words (any number of bits). Word sizes tend to vary depending on the type of microcontroller and the usage of the word; in the PICAXE system, words are 16 bits in length.

OCTAL

The octal system has eight distinct digits: 0, 1, 2, 3, 4, 5, 6, and 7. Individual digits are written next to each other to make up a number and each digit is related to its neighbor by a factor of 8. Octal digits are made up of three binary digits and it is, therefore, easy to convert between binary and octal. Currently, octal numbers are not often used.

HEXADECIMAL

The hexadecimal system has 16 distinct digits: 0, 1, 2, 3, 4, 5, 6, 7, 8, 9, A, B, C, D, E, and F. Individual digits are written next to each other to make up a number and each digit is related to its neighbor by a factor of 16. The digits A to F can be written in upper or lower case. The hexadecimal system is often used to represent binary numbers because each hexadecimal digit represents four binary digits; thus, two hexadecimal digits represent a byte. Hexadecimal digits are sometimes referred to as nibbles and the term upper-nibble refers to the most significant four bits of a byte. Lower-nibble refers to the least significant four bits of a byte.

DECIMAL

The decimal system has 10 distinct digits: 0, 1, 2, 3, 4, 5, 6, 7, 8, and 9. Individual digits are written next to each other to make up a number. If the number has a fractional part, the integer portion and the fractional portion are separated by a decimal point, which is written as a dot (or a comma in some countries). Each decimal digit is related to its neighbor by a factor of 10.

BINARY-CODED DECIMAL

The binary-coded decimal system uses four binary digits to represent a single decimal digit. Only 10 of the possible 16 combinations of four bits are used. Two BCD digits can be stored in a byte, thus giving a number in the range 0–99, although, on some occasions, only one BCD digit is stored in a byte, giving a number in the range 0–9. By contrast, if the byte were to be used as a binary number, there would be 256 different numbers that could be stored (0–255). Thus, the BCD system is wasteful of storage space, but does have advantages in some circumstances.

ASCII

ASCII is a not really a number system but a code that is used for transmitting data. However, it is a common task to convert numbers to and from ASCII; therefore, it is considered here. ASCII is an acronym that stands for American Standard Code for Information Interchange and, while there are other codes in use, ASCII is the code that is probably most often used for data transmission between microcontrollers and/or hosts. ASCII codes consist of eight bits (some implementations of ASCII use only seven bits for data and the eighth bit is used for a parity check) that are stored in a byte. Eight bits allow 256 different combinations and ASCII codes can be represented as a hexadecimal number, decimal number, character, or abbreviations for control characters.

The first 32 ASCII codes are used as control characters, such as carriage return (abbreviated CR) and linefeed (abbreviated LF). The next 95 codes are used for printing characters such as numeric digits, letters, and special characters, code 128 is the delete code, and the remaining 128 codes are generally used as application-specific symbols or not used at all. ASCII codes are shown in the appendix; some commonly used ASCII codes are:

ASCII Code	Use
Hex. 07 (decimal 7)	Bell
Hex. 0A (decimal 10)	Linefeed
Hex. 0D (decimal 12)	Carriage return
Hex. 20 (decimal 32)	Space
Hex 30 to 39 (decimal 48 to 57)	Numeric digits 0–9
Hex. 41 to 5A (decimal 65 to 90)	Uppercase letters A–Z
Hex. 61 to 7A (decimal 97 to 122)	Lowercase letters a–z

Picaxe Arithmetic

PICAXE arithmetic operators perform arithmetic on unsigned integers that are held in bytes or words. Bytes contain binary numbers in the range 0 to 255 and words contain binary numbers in the range 0 to 65,535. Arithmetic operations can operate on combinations of bytes and words.

Programming Editor provides the following six arithmetic operators to perform addition, subtraction, multiplication, and division.

+	Numeric addition
−	Numeric subtraction
*	Numeric multiplication that returns the low portion of the result
**	Numeric multiplication that returns the high portion of the result
/	Numeric division that returns the quotient
//	Numeric division that returns the remainder

When adding or multiplying numbers, overflow will occur when the result exceeds 255 for a byte or 65,535 for a word. When subtracting, underflow will occur if the result becomes less than zero.

Examples of addition and subtraction

```
b0 = 255        'Set b0 to the maximum value for a byte
b0 = b0 + 1     'Add 1 to b0 and place the result in b0
```

In this example, overflow has occurred. The result should be 256, which is too large to store in a byte. Consequently, the result is truncated and b0 will contain 0 (zero).

```
b0 = 255        'Set b0 to the maximum value for a byte
w1 = b0 + 1     'Add 1 to b0 and place the result in a word
```

This example is the same as the previous example, except the result is placed in a word so that no overflow occurs; w1 will contain 256, which is the correct result.

w0 = 65535   'Set w0 to the maximum value for a word
w0 = w0 + 1   'Add 1 to w0 and place the result in w0

In this example, the result should be 65,536 which is too large to store in a word. Consequently, the result is truncated and w0 will contain 0 (zero).

b0 = 0   'Set b0 to the minimum value for a byte
b0 = b0 – 1   'Subtract 1 from b0 and place the result in b0

In this example, underflow has occurred. The result should be –1 (minus one) and b0 contains 255. Some readers will recognize 255 as being –1 (minus one) in two's complement form; this concept is discussed further later in this book.

w0 = 0   'Set b0 to the minimum value for a word
w0 = w0 – 1   'Subtract 1 from w0 and place the result into w0

In this example, underflow has occurred. The result should be –1 (minus one) and w0 contains 65,535. Some readers will recognize 65,535 as being –1 (minus one) in two's complement form. This concept is discussed further later in this book.

MULTIPLICATION

When multiplying numbers, the maximum size of the product is equal to the sum of the size of multiplier and the size of the multiplicand, i.e., when multiplying two bytes, the result requires up to 16 bits. When multiplying a byte and a word, the result requires up to 24 bits. When multiplying two words the result requires up to 32 bits. The largest number the PICAXE can store is 16 bits, which will cause truncation when the result of a multiply operation exceeds 16 bits in size. To overcome this, the PICAXE has two multiply operators, * and **. The * operator returns the low portion of the product, while the ** operator returns the high portion of the product.

Examples of multiplication

b0 = 10
b1 = 20
b2 = b0 * b1   'Multiply b0 and b1, and place the low portion of the result in b2

In this example, two numbers are multiplied and the result is placed in a byte. In practice, a word (16 bits) would be required to hold all possible results from a multiplications of two bytes. However, in this case the result is 200, which can be stored in a byte without any truncation occurring.

b0 = 10
b1 = 30
b2 = b0 * b1   'Multiply b0 and b1, and place the low portion of the result in b2
b3 = b0 ** b1   'Multiply b0 and b1, and place the high portion of the result in b3

In this example, two 8-bit numbers are multiplied and the low portion of the result is placed in a byte (b2). The result of this multiplication is 300, which is too large to store in a byte, so truncation occurs and b2 will contain 44. The same numbers are multiplied again and the high portion of the result is placed in another byte (b3), which will contain 1 after the operation. The word variable w1, which is made up of b3 and b2, will contain 300, which is the correct result. The single multiply command w1 = b0 * b1 could have been used here to produce the same result.

```
w1 = 40000
b0 = 10
w2 = b0 * w1       'Multiply w1 and b0, and place the low portion of the result in w2
w3 = b0 ** w1      'Multiply w1 and b0, and place the low portion of the result in w3
```

In this example, an 8- and a 16-bit number are multiplied and the low portion of the result is placed in w2. Then, the same two numbers are multiplied again and the high portion of the result is placed in w3. The result of this multiplication is 400,000, well in excess of the maximum number that can be stored in a word. After these operations, w2 will contain 6784 and w3 will contain 6.

DIVISION

When dividing numbers, the maximum size of the quotient is equal to the size of the dividend; the maximum size of the remainder is equal to the size of the divisor. Thus, if the dividend is a byte, the quotient will need to be a byte (or a word); if the dividend is a word, the quotient will need to be a word. If the divisor is a word, the remainder will need to be a word; if the divisor is a byte, then the remainder will need to be a byte (or a word). The PICAXE has two division operators, / and //. The / operator returns the quotient and the // operator (modulus divide) returns the remainder.

Examples:

```
w0 = 65534
w1 = 65535
w2 = w0 / w1       'Divide w0 by w1 and place the quotient in w2
w3 = w0 // w1      'Divide (modulus) w0 by w1 and place the remainder in w3
```

In this example, the number 65534 is divided by the number 65535. The resulting quotient in w2 is 0 and the remainder in w3 is 65534.

```
w0 = 65535
w1 = 2
w2 = w0 / w1       'Divide w0 by w1 and place the quotient in w2
w3 = w0 // w1      'Divide (modulus) w0 by w1 and place the remainder in w3
```

In this example, the number 65535 is divided by the number 2. The resulting quotient in w2 is 32,767 and the remainder in w3 is 1.

```
b0 = 254
b1 = 255
b2 = b0 / b1       'Divide b0 by b1 and place the quotient in b2
b3 = b0 // b1      'Divide (modulus) b0 by b1 and place the remainder in b3
```

In this example, the number 254 is divided by the number 255. The resulting quotient in b2 is 0 and the remainder in b3 is 254.

b0 = 255
b1 = 2
b2 = b0 / b1   'Divide b0 by b1 and place the quotient in b2
b3 = b0 // b1   'Divide (modulus) b0 by b1 and place the remainder in b3

In this example, the number 255 is divided by the number 2. The resulting quotient in b2 is 127 and the remainder in b3 is 1.

Boolean Arithmetic

Boolean algebra (named after George Boole, a 19th century mathematician/philosopher) is the mathematics that is used to express logical operations. Boolean values can have one of two states called "true" and "false."

The primary Boolean operations are:

NOT: Logical inversion

AND: Logical AND

OR: Logical OR, sometimes called *Inclusive OR* (IOR)

XOR: Logical Exclusive OR

NAND: Logical NAND

NOR: Logical NOR

Shift left: Logical shift left

Shift right: Logical shift right

Logical NOT takes a single input and inverts it. Logical AND takes two, or more, inputs and if all inputs are true, the output will be true, otherwise the output is false. Logical OR takes two, or more, inputs and if any input is true, the output will be true, otherwise the output is false. Logical XOR takes two inputs and if either input is true the output is true; if all inputs are false or all inputs are true, then the output is false. Logic NAND is equivalent to logic AND followed by Logic NOT; Logic NOR is equivalent to logic OR followed by logic NOT. Shift left moves the bits of a byte or word to the left and shift right moves the bits of a byte or word to the right. Table 5.1 illustrates Boolean operations.

The term "true" also represents the states of "logic high" and binary 1 (one), and "false" the represents logic low, or binary 0 (zero). This equivalency is illustrated in Table 5.2.

Programming Editor provides the logical AND, OR, XOR, NAND, and NOR operators for all PICAXE chips and the shift left and shift right operators for the X1 and X2 chips. The logical NOT function can be synthesized by XOR'ing with 1s, shift left can be synthesized by multiplying by 2, and shift right can be synthesised by dividing by 2.

TABLE 5.1	BOOLEAN TRUTH TABLE					
INPUT	NOT					
False	True					
True	False					
INPUT 1	INPUT 2	AND	OR	XOR	NAND	NOR
False	False	False	False	False	True	True
False	True	False	True	True	True	False
True	False	False	True	True	True	False
True	True	True	True	False	False	False

When using logical operators on bytes or words, the operation regards the byte or word as a series of individual bits. Table 5.3 shows the results of logical operations on bytes; the principles are the same for words.

Examples:

```
'Logical NOT, (XOR with 1's is used to synthesize the logical NOT operation)
b1 = %00111010        'Hex 3A
b2 = %11111111        "Hex FF
b0 = b1 XOR b2        "Result in b0 = %11000101, $C5

'Logical AND
b1 = %00111010        "Hex 3A
b2 = %01010011        "Hex 53
b0 = b1 AND b2        "Result in b0 = %00010010, $12

'Logical OR
b1 = %00111010        "Hex 3A
b2 = %01010011        "Hex 53
b0 = b1 OR b2         "Result in b0 = %01111011, $7B

'Logical XOR
b1 = %00111010        "Hex 3A
b2 = %01010011        "Hex 53
b0 = b1 XOR b2        "Result in b0 = %01101001, $69
```

TABLE 5.2	LOGICAL EQUIVALENCY			
True	→	Logic high	→	Binary 1
False	→	Logic low	→	Binary 0

	BIT 7	BIT 6	BIT 5	BIT 4	BIT 3	BIT 2	BIT 1	BIT 0
TABLE 5.3 LOGICAL OPERATIONS ON BYTES								
Input								
Byte 1	0	0	1	1	1	0	1	0
Byte 2	0	1	0	1	0	0	1	1
	↓	↓	↓	↓	↓	↓	↓	↓
AND	0	0	0	1	0	0	1	0
OR	0	1	1	1	1	0	1	1
XOR	0	1	1	0	1	0	0	1
NAND	1	1	1	0	1	1	0	1
NOR	1	0	0	0	0	1	0	1
	0	0	1	1	1	0	1	0
Shift left	↙	↙	↙	↙	↙	↙	↙	↙
	0	1	1	1	0	1	0	0 ←
	0	1	0	1	0	0	1	1
Shift right	↘	↘	↘	↘	↘	↘	↘	↘
	→ 0	0	1	0	1	0	0	1

```
'Logical NAND
b1 = %00111010        "Hex 3A
b2 = %01010011        "Hex 53
b0 = b1 NAND b2       "Result in b0 = %11101101, $ED


'Logical NOR
b1 = %00111010        "Hex 3A
b2 = %01010011        "Hex 53
b0 = b1 NOR b2        "Result in b0 = %10000100, $84


'Logical NOR
b1 = %00111010        "Hex 3A
b2 = %01010011        "Hex 53
b0 = b1 NOR b2        "Result in b0 = %10000100, $84


'Logical Shift left
b1 = %00111010        "Hex 3A
b1 = b1 * 2           "Result in b1 = %01110100, $74


'Logical shift right
b1 = %01010011        "Hex 53
b1 = b1 / 2           "Result in b1 = %00101001, $29
```

```
'Logical AND using words
w1 = %0011101011110000        "Hex 3AF0
w2 = %0101001101110111        "Hex 5377
w0 = w1 AND w2                "Result in w0 =%0001001001110000, $1270
```

Logical operations are often used for isolating, setting, clearing, inverting, and moving bits.

Some useful rules when using logical operations are:

ANDing a bit with a zero will result in a zero; ANDing a bit with a 1 will result in the original bit.

ORing a bit with a 1 will result in a 1; ORing a bit with a 0 will result in the original bit.

XORing a bit with a 1 will result in the bit being inverted; and XORing a bit with a 0 will result in the original bit.

XORing a byte or word with 1's ($FF for a byte and $FFFF for a word) will result in all bits being inverted (Logical NOT).

Examples:

pins = outpins AND %11111010 'Set pins 2 & 0 to logic low, other pins unchanged

In this example, pin2 and pin0 are set to logic low state at the same time without affecting the state of any other output pins.

pins = outpins OR %00000011 'Set pins 1 and 0 to logic high, other pins unchanged

In this example, pin1 and pin0 are set to logic high state at the same time without affecting the state of any other output pins. This requirement can be met by OR'ing pin1 and pin0 with 1's and all other pins with 0's.

pins = outpins XOR %00010000 'Invert out pin 4, other pins unchanged

In this example, the state of pin4 is inverted without affecting the state of any other output pins. The solution is to XOR pin4 with a 1 and XOR all other pins with 0's.

```
b0 = pins & %00000110                'Isolate pin2 and pin1
b0 = b0 / 2                          'Shift right one bit to make a number 0 to 3
branch b0, (label0, label1, label2, label3) 'Branch depending on the pin state
```

This example shows how to branch to one of four different labels depending on the state of input pin2 and input pin1. The states of pin2 and pin1 are isolated by ANDing them with 1s and the result is shifted right to create a number in the range 0 to 3, which can be used as the offset in a **branch** command.

readtemp12 1, w6   'Read a 12 bit temperature value into w6
b0 = w6 AND $000F   'Put the fractional portion in b0, leaving w6 unchanged
w6 = w6 AND $FFF0 / 16   'Isolate the integer portion and align it in bits 7 to 0
                                                     of w6

This example shows how to isolate the integer and fractional portions of a 12-bit temperature reading that has been read into w6 by the **readtemp12** command. The reading consists of eight integer bits and four fractional bits. The integer and fractional portions are isolated by ANDing them with 1s. The integer portion will still be in bits 11 to 4 after being isolated and will need to be moved to bits 7 to 0 by shifting right four bits, which is achieved by division by 16.

Data Conversion

The different number schemes that are in use mean that there will be occasions when we need to convert data from one number system to another. Commonly used conversions are decimal to binary, decimal to BCD, BCD to ASCII, binary to hexadecimal, binary to BCD, and BCD to ASCII.

Many calculators, including software calculators that are often supplied with PC operating systems, can perform conversions between binary, octal, hexadecimal, and decimal.

CONVERTING BETWEEN DECIMAL DIGITS (BCD) AND ASCII

BCD and single decimal digits are effectively the same except that one is expressed in decimal and the other is expressed in binary. There is a direct mathematical relationship between decimal, BCD, and the ASCII codes for the digits "0" to "9." Converting BCD to ASCII is a simple process of adding hexadecimal 30 (or decimal 48) to the decimal digit. To convert the ASCII characters 0–9 to decimal digits, subtract hexadecimal 30 (or decimal 48) from the ASCII code.

The relationship between decimal, BCD, and ASCII digits 0–9 is shown in the tabulation:

Decimal	BCD	ASCII Code
0	0000	hex. 30 (or decimal 48)
1	0001	hex. 31 (or decimal 49)
2	0010	hex. 32 (or decimal 50)
3	0011	hex. 33 (or decimal 51)
4	0100	hex. 34 (or decimal 52)
5	0101	hex. 35 (or decimal 53)
6	0110	hex. 36 (or decimal 54)
7	0111	hex. 37 (or decimal 55)
8	1000	hex. 38 (or decimal 56)
9	1001	hex. 39 (or decimal 57)

CONVERTING BETWEEN BINARY AND OCTAL

Each octal digit consists of three binary digits and it is, therefore, relatively easy to convert between binary and octal by inspection. Currently, octal is not often used.

The relationship between binary and octal is seen in the tabulation below:

Binary	Octal
000	0
001	1
010	2
011	3
100	4
101	5
110	6
111	7

Example:

Binary 11010001 = Octal 321

CONVERTING BETWEEN BINARY AND HEXADECIMAL

Each hexadecimal digit consists of four binary digits and, with practice, it is relatively easy to convert between binary and hexadecimal by inspection. Hexadecimal is commonly used when working with microcontrollers.

The relationship between binary and hexadecimal is shown in the tabulation below:

Binary	Hexadecimal
0000	0
0001	1
0010	2
0011	3
0100	4
0101	5
0110	6
0111	7
1000	8
1001	9
1010	A
1011	B
1100	C
1101	D
1110	E
1111	F

Example:

Binary 11010001 = Hexadecimal D1

CONVERTING BINARY TO DECIMAL AND BCD

In principle, to convert binary integers to decimal the method is to divide the binary number by 10 and isolate the remainder. The remainder is the least significant decimal digit (and it is also a BCD digit). This process is repeated until all the digits have been processed. Note that the least significant digit will be isolated first.

Consider the following code segment that converts a single byte binary number to ASCII and sends each digit to the serial port commencing with the least significant digit.

b0 = 123	'Choose a binary number, in this case 123
do	
b1 = b0 // 10	'Divide the binary number by 10 and place the remainder in b1
b1 = b1 + 48	'Convert to ASCII
b0 = b0 / 10	'Divide the binary number by 10 and save the quotient in b0
serout 1, T2400, (b1)	'Send ASCII character to a serial port
loop while b0 <> 0	'If the binary number is not zero, repeat the process

Code analysis:

b0 = 123	A binary number is chosen. It can be any number in the range 0 to 255. In this case, it is 123.
b1 = b0 // 10	The binary number is divided by 10 and the remainder is placed into variable b1. The remainder is a bcd digit and, in this example will be equal to 3 on the first iteration of the loop, 2 on the second iteration, and 1 on the third iteration.
b1 = b1 + 48	The decimal number 48 is added to convert the bcd digit to an ASCII character. Note that this and the previous command could be combined into the single command b1 = b0 // 10 + 48.
b0 = b0 / 10	The original number is divided by 10 and the quotient is retained in b0.
serout 1, T2400, (b1)	This command sends the ASCII character in b1 to a serial port. On the first iteration of the loop, "3" will be sent, on the second "2" will be sent, and on the third "1" will be sent. Note also that the BCD-to-ASCII conversion could have been performed here with the command serout 1, T2400, (#b1).
if b0 <> 0 then nextdigit	The result is compared to 0 and, if it is 0, then no more significant decimal digits remain to be converted and the process ends. If b0 is not yet 0, then there are still more digits to be converted.

The preceding code will output between one and three ASCII characters depending on the value of the number to be converted. If the original binary number was 1, then only one ASCII character is output; if it was 45, then two ASCII characters will be output. In some circumstances, it may be desirable to output a fixed number of characters irrespective of the value of the original number. This can be achieved by using a **for ... next** loop.

Example:

```
b0 = 45                  'Choose a binary number, in this case 45
for b9 = 1 to 3          'Set up a loop that repeats three times
    b1 = b0 // 10 + 48   'Isolate the bcd digit and convert to ASCII
    b0 = b0 / 10         'Divide the binary number by 10
    serout 1, T2400, (b1) 'Send the ASCII digit to the serial port
next b9                  'Repeat until three digits are processed
```

In this example, three ASCII digits are always sent to the serial port. They will be "5" on the first iteration, "4" on the second iteration, and "0" on the third iteration.

This code can be enhanced to replace leading zeros with spaces.

```
b0 = 45                  'Choose a binary number, in this case 45
for b9 = 1 to 3          'Set up a loop that repeats three times
    b1 = b0 // 10 + 48   'Isolate the bcd digit and convert to ASCII
    b0 = b0 / 10         'Divide the binary number by 10
    if b0 = 0 and b1 = "0" then 'If there are no more significant digits and
                                'current digit is "0"
        b1 = " "         'then change the "0" to a space
    endif
    serout 1, T2400, (b1) 'Send the ASCII digit to the serial port
next b9                  'Repeat until three digits are processed
```

In this example, three ASCII digits are always sent to the serial port, however, leading 0s are replaced with spaces. The digits sent to the serial port will be "5" on the first iteration, "4" on the second iteration, and a space on the third iteration.

The **bintoascii** and **serout** commands can also perform conversion between binary and ASCII. The **bintoascii** command takes a byte or word variable and places the resulting ASCII characters in 3- or 5-byte variables depending on whether a byte or word was converted. The **serout** command has the capability to convert from binary to ASCII during the output process if the "#" operator is specified in front of a byte or word variable.

Examples:

```
b0 = 123                 'Set b0 to the number 123
bintoascii b0, b1, b2, b3 'Convert b0 to ASCII characters in b1, b2, & b3
```

In this example, the number 123 is placed in a byte and then converted to ASCII in b1, b2, b3. The result is that b1 will contain the ASCII code for 1 (hundreds digit), b2 will contain ASCII 2 (tens digit), and b3 will contain ASCII 3 (units digit). Only three characters are possible when converting a byte, because the largest number that can be held in a byte is 255.

w0 = 45678   'Set w0 to the number 45,678
bintoascii w0, b2, b3, b4, b5, b6   'Convert w0 to ASCII characters in b2, b3,
                                           b4, b5, and b6

In this example, the number 45, 678 is placed in a word and then converted to ASCII in b2, b3, b4, b5, and b6. The result is that b2 will contain the ASCII code for 4 (ten thousands digit), b3 will contain ASCII 5 (thousands digit), b3 will contain ASCII 6 (hundreds digit), b4 will contain ASCII 7 (tens digit), and b6 will contain ASCII 8 (units digit). Five characters are needed when converting a word because the largest number that can be held in a word is 65,535.

b0 = 234   'Set b0 to two hundred and thirty four
serout 1, T2400, (#b0)   'Send b0 to a serial port and convert to ASCII

In this example, the number two hundred and thirty four is placed in a byte and then transmitted to a serial port. The "#" operator is specified in front of b0 and b0 is converted to ASCII for transmission. The result is that the ASCII characters "2," "3," and "4" will be sent to the serial port.

The **bcdtoascii** command will convert two bcd digits held in a byte, or four bcd digits held in a word, to ASCII characters.

Examples:

b0 = $45   'Set b0 to the bcd digits 4 and 5
bintoascii b0, b1, b2   'Convert b0 from bcd to ASCII characters in b1, b2

In this example, the bcd digits 4 and 5 are placed in a byte and then converted to ASCII in b1 and b2. The result is that b1 will contain the ASCII code for 4 (tens digit), and b2 will contain ASCII 5 (units digit). Only two characters are possible when converting a byte because the largest number that can be held in a byte using bcd is 99.

w0 = $6789   'Set w0 to the bcd digits 6, 7, 8 , and 9
bintoascii w0, b2, b3, b4, b5   'Convert w0 from bcd to ASCII characters in b2, b3,
                                           b4, and b5

In this example, the bcd digits 6, 7, 8 and 9 are placed in a word and then converted to ASCII in b2, b3, b4, b5. The result is that b2 will contain the ASCII code for 6 (thousands digit), b3 will contain ASCII 7 (hundreds digit), b4 will contain ASCII 8 (tens digit), and b5 will contain ASCII 9 (units digit). Four characters are possible when converting a word because the largest number that can be held in a word using bcd is 9999.

CONVERTING ASCII, BCD, AND DECIMAL TO BINARY

When ASCII digits are to be used in mathematical calculations, they must first be converted to BCD (decimal) digits and then to binary numbers. Separate processes are required for integers and fractions and the conversion process is relatively simple once it is understood.

ASCII Integers

The principle is to convert each ASCII digit to a BCD digit, then scale the decimal digits by a weighting factor, and add them to a binary number that has been initialized to zero. If the ASCII number has a sign, then once the conversion is complete, test the sign for negative and complement the result.

Consider the following code segment, which converts three ASCII digits to a single-byte binary number. The ASCII digits are read from the serial port, least significant digit first.

```
b0 = 0                  'Initialize the binary number to zero
b1 = 1                  'Initialize the weighting factor
for b3 = 1 to 3         'Set up a loop that repeats three times
   serin 1, T2400, b4   'Read the next ASCII digit from the serial port
   b4 = b4 – 48         'Convert the ASCII digit to a BCD digit
   b4 = b4 * b1         'Apply the weighting factor
   b0 = b0 + b4         'Add the scaled digit to the binary number
   b1 = b1 * 10         'Scale the weighting factor for the next digit
next b3                 'Repeat until all digits are processed
                        'Complement b0 here if the sign was negative
```

Code analysis

b0 = 0	The variable b0 will be used to hold the converted binary number and must be initialized to 0. It is a byte variable and can hold numbers in the range 0 to 255. If larger numbers are required, a word variable should be used.
b1 = 1	The variable b1 is used to hold the weighting factor and is initialized to 1 for the least significant digit. It will be scaled up by a factor of 10 for each following digit.
for b3 = 1 to 3	A loop is set up to repeat three times, once for each digit.
serin 1, T2400, b4	This command reads the ASCII digits from the serial port, one digit at a time, least significant digit first.
b4 = b4 – 48	This command converts the ASCII digit to BCD by subtracting 48. Note that this command and the previous command could be replaced by the single command serin 1, T2400, #b4.
b4=_ b4 * b1	This command applies the weighting factor according to the digit weight. For the least significant digit, the weighting factor is 1, for the second digit it is 10, and for the third digit it is 100.
b0 = b0 + b4	Add the scaled digit to the binary number.

b1 = b1 * 10	Multiply the weighting factor by 10 in readiness for the next digit.
next b3	Repeat until three digits have been processed.
	For signed ASCII numbers, test the sign at this point and complement the binary number (b0) if the sign is negative.

Note that three ASCII digits can hold numbers in the range 000 to 999 and that a single byte can hold a maximum of number of 255. Thus, it is possible to exceed the capacity of a byte with three ASCII digits. In this example, it is presumed that the number will not exceed 255. If the ASCII number to be converted will exceed 255, then a word variable that can hold numbers up to 65,535 should be used for the converted binary number. Note also that if ASCII digits are received most significant digit first, then the scaling factor should be initialized to the highest value and divided by 10, as in the following example, which converts four ASCII digits to a word variable, with the most significant digit being received first.

w0 = 0	'Initialize a word to zero
w1 = 1000	'Initialize the weighting factor - must be a word to hold 1000
for b4 = 1 to 4	'Set up a loop that repeats four times
serin 1, T2400, #b5	'Read the ASCII digits and convert to bcd
w3 = b5 * w1	'Apply the weighting factor
w0 = w0 + w3	'Add the scaled digit to the binary number
w1 = w1 / 10	'Decrease the scaling factor for the next digit
next b4	'Repeat until four digits have been processed

ASCII Fractions

To process ASCII fractions, first separate the fraction from any integer part of the number, and then process the fractional digits as if they were an integer. To recombine the integer part and the fractional part, multiply the integer part by a factor of 10 for each decimal place in the fractional part, and then add the scaled integer and fractional parts together. This will produce a scaled decimal number; producing binary fractions is discussed later in this chapter.

Consider the following code segment, which converts a six-digit ASCII number with three integer digits, a decimal point, and two fractional digits, into a word variable. The ASCII digits are read from the serial port, least significant digit first.

	'Fractional portion
b0 = 0	'Initialize the fraction to zero
b1 = 1	'Initialize the weighting factor
for b3 = 1 to 2	'Set up a loop that repeats twice, once for each digit
serin 1, T2400, b4	'Read the next ASCII fraction digit from the serial 'port
b4 = b4 – 48	'Convert the ASCII digit to a BCD digit
b4 = b4 * b1	'Apply the weighting factor
b0 = b0 + b4	'Add the scaled digit
b1 = b1 * 10	'Scale the weighting factor for the next digit

```
next b3                  'Repeat until all digits are processed
                         'The fraction multiplied by 100 is now in b0
serin 1, T2400, b4       'Read the decimal point and discard it
                         'Integer portion
w3 = 0                   'Initialize the integer to zero
b1 = 1                   'Initialize the weighting factor
for b3 = 1 to 3          'Set up a loop that repeats three times, once for each digit
   serin 1, T2400, b4    'Read the next ASCII digit from the serial port
   b4 = b4 – 48          'Convert the ASCII digit to a BCD digit
   b4 = b4 * b1          'Apply the weighting factor
   w3 = w3 + b4          'Add the scaled digit
   b1 = b1 * 10          'Scale the weighting factor for the next digit
next b3                  'Repeat until all digits are processed
                         'The integer is now in w3
w3 = w3 * 100 + b0       'Combine the integer and fraction
                         'The maximum number that can be held is 655.35
                         'Note that this is a decimal number scaled up by a factor
                         'of 100, not a binary fraction.
```

Advanced Arithmetic

All PICAXE arithmetic is performed with integers; however, with appropriate care and program code, it is possible to perform arithmetic on decimal fractions, signed binary numbers, binary fractions, and BCD numbers.

DECIMAL FRACTIONS

Arithmetic can be performed on decimal fractions by scaling them by powers of 10 so that they are in the range for integers, e.g., the decimal number 12.3 would be multiplied by 10 and stored as 123.

ADDING AND SUBTRACTING DECIMAL FRACTIONS

When adding or subtracting fractions, each number must have the same number of decimal places.
   Example:

```
w1 = 1953        'This is the number 19.53 scaled up by a factor of 100
w2 = 210         'This is the number 2.1 scaled up by a factor of 100
w3 = w1 + w2     'Result = 2163 which is 21.63 scaled up by a factor of 100
```

MULTIPLYING DECIMAL FRACTIONS

When multiplying fractions, each number can have any number of decimal places and the number of decimal places in the result will be equal to the sum of the number of decimal places in the multiplier and multiplicand.

Example:

w1 = 1953	'This is the number 19.53 scaled up by a factor of 100
w2 = 21	'This is the number 2.1 scaled up by a factor of 10
w3 = w1 * w2	'Result = 41013 which is 41.013 scaled up by a factor of 1000

DIVIDING FRACTIONS

When dividing fractions, each number can have any number of decimal places. The number of decimal places in the quotient is equal to the number of decimal places in the dividend minus the number of decimal places in the divisor. If the number of decimal places in the quotient is a negative number, then the result needs to be multiplied by powers of 10 to obtain the correct result.

Example:

w1 = 1953	'This is the number 19.53 scaled up by a factor of 100
w2 = 21	'This is the number 2.1 scaled up by a factor of 10
w3 = w1 / w2	'Result = 93 which is 9.3 scaled up by a factor of 10

CONVERTING DECIMAL FRACTIONS TO ASCII

The process of converting decimal fractions is essentially the same as converting decimal numbers to ASCII except that the decimal point must be inserted at the appropriate place. Note that the **bintoascii** and "#" operator in the **serout** command can be used if the decimal point is not required.

Example:

b0 = 93	'Choose a binary number, in this case 9.3
b1 = b0 // 10 + 48	'Isolate ls digit and convert to ASCII
serout 1, T2400, (b1)	'Send ASCII character to serial port
serout 1, T2400, (".")	'Send decimal point to serial port
b1 = b0 / 10	'Isolate ms digit and convert to ASCII
serout 1, T2400, (b1)	'Send ASCII character to a serial port

BCD ARITHMETIC

The nature of BCD makes it relatively easy to develop software routines to perform addition and subtraction on it. In practice, two BCD digits can be stored in 1 byte. However, these examples are given with 1 digit per byte. BCD arithmetic with two digits per byte is used the LCD clock/calendar experiment later in this book.

BCD ADDITION

The PICAXE addition operator (+) can be used to add two bytes containing bcd digits. If the result is 10 or more then a carry has occurred and it is necessary to subtract 10 from the result and set a carry indicator.

Example:

b0 = 1	'Set up the bcd number 123 in b0, b1, and b2 using one digit per byte
b1 = 2	
b2 = 3	
b3 = 4	'Set up the bcd number 468 in b3, b4, and b5 using one digit per byte
b4 = 6	
b5 = 8	
b2 = b2 + b5	'Add least significant digits (units) and place the result in b2
if b2 >= 10 then	'Test for carry
b2 = b2 – 10	'Carry has occurred, subtract 10 and
b6 = 1	'set carry indicator
else b6 = 0	'Carry has not occurred, clear carry indicator
endif	
b1 = b1 + b4	'Add the next significant digits (tens) and place the result in b1
b1 = b1 + b6	'Add the carry from the previous digit
if b1 >= 10 then	'Test for carry
b1 = b1 – 10	'Carry has occurred, subtract 10 and
b6 = 1	'set carry indicator
else b6 = 0	'Carry has not occurred, clear carry indicator
endif	
b0 = b0 +b3	'Add the most significant digits (hundreds) and place the result in b0
b0 = b0 + b6	'Add the carry from the previous digit
if b0 >= 10 then	'Test for carry
b0 = b0 – 10	'Carry has occurred, subtract 10 and
b6 = 1	'set carry indicator
else b6 = 0	'Carry has not occurred, reset carry indicator
endif	

The result is that the bcd number 591 is now in b0, b1, and b2 and a carry of 0 (zero) is in b6. In this example, the variable b6 is used as a carry indicator and is added to the result of the addition of the tens digit (where carry occurred from the units digits) and the result of the addition of the hundreds digit (where carry did not occur). If, at the end of the addition, the carry indicator was set to 1 (one), then a carry would have occurred and a fourth bcd digit would be necessary to store the correct result. While this procedure seems unnecessarily complex, it does have the advantages that numbers greater than 65,535 can be processed and a carry indicator is available. It is also possible to set this code up as a loop using pointers.

The contents of the variables b0 to b6 after each command are shown in Table 5.4.

SUBTRACTION OF BCD DIGITS

The PICAXE subtraction operator (−) can be used to subtract two bytes containing bcd digits. If the result is less than 0, then borrow has occurred and it is necessary to add

TABLE 5.4   WORKING FOR BCD ADDITION EXAMPLE

COMMAND	B0	B1	B2	B3	B4	B5	B6
b0 = 1	1	x	x	X	X	x	X
b1 = 2	1	2	x	X	X	x	X
b2 = 3	1	2	3	X	X	x	X
b3 = 4	1	2	3	4	X	x	X
b4 = 6	1	2	3	4	6	x	X
b5 = 8	1	2	3	4	6	8	X
b2 = b2 + b5	1	2	11	4	6	8	X
if b2 > 10 then	1	2	11	4	6	8	X
b2 = b2 − 10	1	2	1	4	6	8	X
b6 = 1	1	2	1	4	6	8	1
else							
b6 = 0							
endif							
b1 = b1 + b4	1	8	1	4	6	8	1
b1 = b1 + b6	1	9	1	4	6	8	1
if b1 >= 10 then	1	9	1	4	6	8	1
b1 = b1 − 10							
b6 = 1							
else							
b6 = 0	1	9	1	4	6	8	0
endif							
b0 = b0 +b3	5	9	1	4	6	8	0
b0 = b0 + b6	5	9	1	4	6	8	0
if b0 >= 10 then	5	9	1	4	6	8	0
b0 = b0 − 10							
b6 = 1							
else							
b6 = 0	5	9	1	4	6	8	0
endif							

10 to the result and set a borrow indicator. To determine if a borrow has occurred, it is necessary to test the result to determine if it is in the range for a bcd digit, which is 0 to 9. This can be determined by testing the digit for being greater than or equal to 10.

Example:

b0 = 5   'Set up the bcd number 591 in b0, b1, and b2 using one digit per byte

b1 = 9

b2 = 1

b3 = 4   'Set up the bcd number 468 in b0, b1, and b2 using one digit per byte

b4 = 6

b5 = 8

b2 = b2 − b5   'Subtract least significant digits (units) and place the result in b2

```
if b2 >= 10 then      'Test for borrow
    b2 = b2 + 10      'Borrow has occurred, add 10 and
    b6 = 1            'set borrow indicator
  else
    b6 = 0            'Borrow has not occurred, clear the borrow indicator
endif
b1 = b1 – b4          'Subtract the next significant digits (tens) and place the result in
                       b1
b1 = b1 – b6          'Subtract the borrow from the previous digit
if b1 >= 10 then      'Test for borrow
    b1 = b1 – 10      'Borrow has occurred, add 10 and
    b6 = 1            'set borrow indicator
else b6 = 0           'Borrow has not occurred, clear the borrow indicator
endif
b0 = b0 – b3          'Subtract the most significant digits (hundreds) and place the
                       result in b0
b0 = b0 – b6          'Subtract the borrow from the previous digit
if b0 >= 10 then      'Test for borrow
    b0 = b0 – 10      'Borrow has occurred, add 10 and
    b6 = 1            'set borrow indicator
else
    b6 = 0            'Borrow has not occurred, clear the borrow indicator
endif
```

The result is that the BCD number 123 is now in b0, b1, and b2 and a borrow of 0 (zero) is in b6. In this example, the variable b6 is used as a borrow indicator and b6 is subtracted from the result of the subtraction of the tens digit (where borrow occurred from the units digits), and the result of the subtraction of the hundreds digit (where borrow did not occur). If, at the end of the subtraction, the borrow indicator was set to 1 (one), then the result would be a negative number. While this procedure seems unnecessarily complex, it does have the advantages that numbers greater than 65,535 can be processed and a borrow indicator is available. It is also possible to set this code up as a loop using pointers.

The contents of the variables b0 to b6 after each command are shown in Table 5.5.

Arithmetic with Signed Numbers

Signed binary arithmetic is performed using the two's complement number scheme and the PICAXE addition and subtraction operators will correctly produce two's complement results provided that bytes and words are not mixed in expressions. The high-order bit position (bit 7 for a byte and bit 15 for a word) is the sign bit and will be a 1 for negative numbers and a 0 (zero) for positive numbers. In the two's complement system, bytes can contain numbers in the range -128 to $+127$ and words can contain numbers in the range $-32,768$ to $+32,767$.

TABLE 5.5   WORKING FOR BCD SUBTRACTION EXAMPLE

COMMAND	B0	B1	B2	B3	B4	B5	B6
b0 = 5	5	x	x	x	x	x	X
b1 = 9	5	9	x	x	x	x	X
b2 = 1	5	9	1	x	x	x	X
b3 = 4	5	9	1	4	x	x	X
b4 = 6	5	9	1	4	6	x	X
b5 = 8	5	9	1	4	6	8	X
b2 = b2 − b5	5	9	248	4	6	8	X
if b2 >= 10 then	5	9	248	4	6	8	X
b2 = b2 + 10	5	9	3	4	6	8	X
b6 = 1	5	9	3	4	6	8	1
else							
b6 = 0							
endif							
b1 = b1 − b4	5	3	3	4	6	8	1
b1 = b1 − b6	5	2	3	4	6	8	1
if b1 >= 10 then	5	2	3	4	6	8	1
b1 = b1 − 10							
b6 = 1							
else							
b6 = 0	5	2	3	4	6	8	0
endif							
b0 = b0 − b3	1	2	3	4	6	8	0
b0 = b0 − b6	1	2	3	4	6	8	0
if b0 >= 10 then	1	2	3	4	6	8	0
b0 = b0 − 10							
b6 = 1							
else							
b6 = 0	1	2	3	4	6	8	0
endif							

When adding signed binary numbers, overflow will occur when the result exceeds 127 for a byte or 32,767 for a word. When subtracting signed numbers, underflow will occur when the result is less than −128 for a byte and less than −32,768 for a word.

ADDITION AND SUBTRACTION OF SIGNED BINARY INTEGERS

Addition and subtraction of signed binary numbers will produce correct results providing words and bytes are not mixed and number ranges are not exceeded.

Examples:

b0 = −127   'Set b0 to −127 (minus 127 = 10000001 in binary)
b1 = 127   'Set b1 to 127 (plus 127 = 01111111 in binary)
b2 = b0 + b1   'Add b0 to b1 and place the result in b2

The result of this addition is the 9-bit binary number 1 0000 0000. By placing the result in a byte, truncation occurs and b2 contains 0000 0000 in binary, which is 0 (zero) in decimal. If the result had been placed in a word, truncation would not occur and the result would not be correct.

w0 = −400	'Set w0 to −400 (minus 400 = 1111 1110 0111 0000 in binary)
w1 = 600	'Set w1 to 600 (plus 600 = 0000 0010 0101 1000 in binary)
w2 = w0 + w1	'Add w0 to w1 and place the result in w2

The result of this addition is the 17-bit binary number 1 0000 0000 1100 1000. Because a word can only hold 16 bits, truncation occurs and the result in w2 is 0000 0000 1100 1000, which is 200 in decimal.

MULTIPLYING AND DIVIDING SIGNED BINARY NUMBERS

Multiplication and division of negative numbers can be performed with appropriate software. First, convert negative numbers to unsigned positive numbers. Next, perform the multiplication or division, and, finally, restore the sign of the multiplicand or quotient and/or remainder. (If either number was negative, then the result is negative. If both numbers were positive or both numbers were negative, then the result is positive).

A negative number can be converted to a positive number by subtracting the negative number from the maximum number, which can be held in the storage size allocated (i.e., 255 for a byte or 65,535 for a word) and adding 1, or by taking the 1's complement and adding 1. The 1's complement is calculated by exclusive or'ing (XOR) with 1's (bitwise complementing).

Example:

w0 = −20	'Place minus 20 in w0
w1 = 30	'Place plus 30 in w1
w0 = w0 xor %1111111111111111 + 1	'Convert from negative to positive
w2 = w0 * w1	'Multiply positive numbers
w2 = w2 xor %1111111111111111 + 1	'Convert from positive to negative

The command w0 = 65535 − w0 + 1 could have been used instead of w0 = w0 xor %1111111111111111 + 1.

The working for this example is shown below in binary and decimal. Spaces are shown between each group of four bits for clarity.

w0 = −20	w0 = 1111 1111 1110 1100 = −20
w1 = 30	w1 = 0000 0000 0001 1110 = 30
w0 = w0 xor %1111111111111111 + 1	w0 = 0000 0000 0001 0011 (xor) = 19
	w0 = 0000 0000 0001 0100 (+ 1) = 20
w2 = w0 * w1	w2 = 0000 0000 0011 1100 = 60
w2 xor %1111111111111111 + 1	w2 = 1111 1111 1100 0011 (xor) = − 61
	w2 = 1111 1111 1100 0100 (+ 1) = − 60

For the command w0 = 65535 − w0 + 1 the working is:

$$= 1111\ 1111\ 1111\ 1111 = 65{,}535$$
$$= 1111\ 1111\ 1110\ 1100 = -20$$
65,535 − w0 + 1
$$= 0000\ 0000\ 0001\ 0011 = (65{,}535 - w0) = 19$$
$$= 0000\ 0000\ 0001\ 0100 = (+1) = 20$$

The same rules can be applied to division.

COMPARING SIGNED NUMBERS

The comparison operators < (less than) and > (greater than) treat all numbers as being unsigned and comparing a negative number with a positive number will result in the negative number being reported as greater than the positive number. For the purposes of determining if a signed number is greater than or less than another signed number, the numbers should be scaled to place them in the positive domain. Signed byte variables are scaled by adding 128 and signed word variables are scaled by adding 32,768. When signed numbers are scaled in this fashion, they will compare correctly. Signed numbers can be compared for equality without scaling, although bytes and words cannot be mixed in a comparison.

CONVERTING SIGNED BINARY NUMBERS TO DECIMAL

The principle is to save the sign, then complement negative numbers and treat them as positive numbers.

Consider the following code segments that test the sign bits and complement negative numbers held in words and bytes:

Negative numbers held in bytes

```
b1 = −1                 'Negative number
b0 = b1                 'Copy to b0 where bits can be tested
if bit7 = 0 then        'Test sign
   b13 = "+"            'Set sign = positive
   else
   b13 = "−"            'Set sign = negative
   b1 = 255 − b1 + 1    'Complement
endif
```

Negative numbers held in words

```
w1 = −1                 'Negative number
w0 = w1                 'Copy to w0 where bits can be tested
if bit15 = 0 then       'Test sign
   b13 = "+"            'Set sign = positive
   else
   b13 = "−"            'Set sign = negative
   w1 = 65535 − w1 + 1  'Complement
endif
```

Code analysis

w1 = −1	'w1 holds the negative number to be converted, in this case −1
w0 = w1	'Copy the number to w0 where bits can be tested
if bit15 = 0 then	'Test the sign bit, which is in the high-order position of the word (bit 15 is 0 for 'positive and 1 for negative)
b13 = "+"	"Use b13 to hold the sign character, in this case "+"
else	
b13 = "−"	'Set sign character = "−"
w1 = 65535 | w1 + 1	'Complement the number
endif	'The sign character (+ or −) is now in b13 and the number in w1 is a positive number 'that can be converted to ASCII

Arithmetic with Binary Fractions

Binary numbers can have a fractional part to the right of a binary point just as decimal numbers have a fractional part to the right of a decimal point. Note that binary fractions are not the same as decimal fractions that have been scaled by a factors of 10 as discussed above.

Addition, subtraction, multiplication, and division of binary fractions can all be performed by using the same rules as would be used for decimal fractions, i.e., for addition and subtraction, each number must have the same number of binary places. For multiplication, each number can have any number of binary places and the number of binary places in the result is equal to the sum of the number of binary places in the multiplier and the number of binary places in the multiplicand. For division, each number can have any number of binary places and the number of binary places in the quotient is equal to the number of binary places in the dividend minus the number of binary places in the divisor. If the number of binary places in the quotient is a negative number, then the result needs to be multiplied by powers of 2 to obtain the correct result.

CONVERTING BINARY FRACTIONS TO DECIMAL

To convert binary fractions to decimal, it is necessary to separate the integer portion from the fractional portion and convert each portion separately. Converting binary integers to decimal, bcd, and ASCII has been discussed above.

For binary fractions, the principle is to multiply the fraction by 10 and isolate the high-order four bits which will contain the decimal digit. To convert this digit to ASCII, add decimal 48 or hexadecimal 30. The process is then repeated until the required number of digits has been processed. Note that the most significant digit will be isolated first.

Consider the following code segment that converts a four-bit binary fraction in the low-order bits of a byte to ASCII and sends each to the serial port, commencing with the most significant digit.

```
b0 = %00001001          'b0 contains the binary fraction to be converted to ASCII
do
    b0 = b0 * 10        'Multiply the binary fraction by 10
    b1 = b0 & $F0       'Isolate the BCD digit in the high-order bits of b1
    b1 = b1 /16 + 48    'Shift right to the low-order four bits and convert to ASCII
    serout 1, T2400, (b1)   'Send ASCII character to a serial port
    b0 = b0 & $0F       'Mask off high-order bits
loop while b0 <> 0      'If the binary fraction is not zero, repeat the process
```

Code analysis

b0 + %00001001	Choose a binary fraction, in this case 0.5625, and put it in the low-order four bits of b0. The implied binary point is between bit 4 and bit 3 = 0000.1001
do	Begin a do . . . loop command
b0 = b0 * 10	Multiply the binary fraction by 10 which will put the first digit in the high-order four bits of b0 and the remaining fraction in the low-order four bits.
b1 = b0 & $F0	Copy the first digit to b1 and mask off the low-order four bits that contain the remaining fraction. Note that it is not strictly necessary to mask off the low-order four bits, because they will be removed by the shift-right operation in the next command.
b1 = b1 /16 + 48	Move the high-order 4 bits of b1 to the low-order four bits by shifting right 4 bits and then add 48 to convert to ASCII. Note that this command and the previous command could be replaced by the single command b1 = b0 /16 + 48.
serout 1, T2400, (b1)	Send the ASCII character to a serial port. Note that the most significant digit is sent first. The first iteration of the loop will send 5, the second will send 6, the third 2, and the last will send 5. Note also that the BCD-to-ASCII conversion could have been performed here with the command serout 1, T2400, (#b1).
b0 = b0 & $0F	Now mask off the high-order four bits of b0 that still contain the BCD digit that was just processed.
loop until b0 <> 0	If the binary fraction is not zero, repeat the process. Note that this will cause a varying number of digits to be output depending on the value of the fraction. If a fixed number of digits is required, then a **for . . . next** loop can be used.

CONVERTING ASCII FRACTIONS TO BINARY FRACTIONS

To convert ASCII fractions to decimal, it is necessary to separate the integer portion from the fractional portion and convert each portion separately. Converting binary integers to decimal, bcd, and ASCII has been discussed above.

For ASCII fractions, the principle is to start with the least significant digit, convert it to bcd, place it in the high-order four bits of a byte, and divide by 10. The low-order four bits of the byte will contain the binary fraction. The process is then repeated until the required number of digits has been processed.

Consider the following code segment that converts a four-digit ASCII fraction to a binary fraction. The fraction is read from a serial port least significant digit first.

```
b0 = 0                  'Clear b0 which will contain the converted binary fraction
for b6 = 1 to 4         'For four digits
    serin 1, T2400, b1  'Read the next least significant digit
    b1 = b1 – $30       'Convert to bcd
    b1 = b1 * 16        'Shift right four bits to place in high-order four bits
    b0 = b0 + b1        'Add it to the binary fraction
    b0 = b0 / 10        'Divide by 10
next b6                 'Next digit
                        'Resulting binary fraction is now in low-order four bits of b0
```

Code analysis

The fraction chosen for this analysis is 0.5625 and the content for the variables after each command is given in Table 5.6.

Floating-Point Arithmetic

Floating-point arithmetic can be performed by using a floating-point co-processor, such as the uMFPUV2 or uMFPUV3.1. Each of these co-processors can be connected to a PICAXE via I2C or SPI interface and perform floating-point and integer arithmetic with up to 32-bit resolution. More information can be found at the Micromega Corporation website http://micromegacorp.com/.

Lookup

On occasions, the resulting data of a conversion bears no mathematical relationship to the source data. In this case, the **lookup** command can convert a number into some other form of data, such an ASCII character.

The **lookup** command requires a source variable, a destination variable, and a list of items. The source variable is used to look up an item in a list of items and place it into the destination variable.

The syntax of the lookup command is:

lookup source variable, (item0, item1, item2 . . . itemn), destination variable

TABLE 5.6   WORKING FOR ASCII FRACTIONS TO BINARY FRACTIONS EXAMPLE

COMMAND	B2	B1	B0
b0 = 0	X	x	%00000000
for b2 = 1 to 4	1	x	%00000000
serin 1, T2400, b1	1	"5" $35 %00110101	%00000000
b1 = b1 – $30	1	%00000101	%00000000
b1 = b1 * 16	1	%01010000	%00000000
b0 = b0 + b1	1	%01010000	%01010000
b0 = b0 / 10	1	%01010000	%00001000
next b6	2	%01010000	%00001000
serin 1, T2400, b1	2	"2" $32 %00110010	%00001000
b1 = b1 – $30	2	%00000010	%00001000
b1 = b1 * 16	2	%00100000	%00001000
b0 = b0 + b1	2	%00100000	%00101000
b0 = b0 / 10	2	%00100000	%00000100
next b6	3	%00100000	%00000100
serin 1, T2400, b1	3	"6" $36 %00110110	%00000100
b1 = b1 – $30	3	%00000110	%00000100
b1 = b1 * 16	3	%01100000	%00000100
b0 = b0 + b1	3	%01100000	%01100100
b0 = b0 / 10	3	%01100000	%00001010
next b6	4	%01100000	%00001010
serin 1, T2400, b1	4	"5" $35 %00110101	%00001010
b1 = b1 – $30	4	%00000101	%00001010
b1 = b1 * 16	4	%01010000	%00001010
b0 = b0 + b1	4	%01010000	%01011010
b0 = b0 / 10	4	%01010000	%00001001
next b6	5	%01010000	%00001001

The converted binary fraction is now in the low-order four bits of b0. The implied binary point is between bit 4 and bit 3 and with the binary point shown b0 = %0000.1001
The binary weight of the first binary place is $1/2^2 = 0.5$, for the 4th binary place is $1/2^4 = 0.0625$. Adding these together = 0.5625, which is the desired result.

where

source variable contains the position of the item in the list (beginning at 0). If *source variable* is greater than the number of items in the list, then the destination variable will remain unchanged.

item0, *item1*, *item2*, and *itemn* are items in the list.

destination variable will receive one of the items from the list.

Examples:

b0 = 2

lookup b0, ("ABCDEF"), b1

In this example, "C" is placed in b1 because it is in position 2 of the list (lists start at position 0).

b0 = 0

lookup b0, (41, 42, 42, 44, 45), b1

In this example, the number 41 is placed in b1 because it is in position 0 of the list.

Lookdown

The **lookdown** command can convert an ASCII character into a number. The item in the source variable will be compared with items in a list; if a match is found, then the position of the item in the list is placed into the destination variable.

The syntax of the **lookdown** command is

lookdown source variable, (item0, item1, item2, . . . itemn), destination variable

where

source variable contains the value to be compared to the list items. If the *source variable* is not found in the list, then the destination variable will be unchanged.

item0, item1, item2, and *itemn* are the list of items.

destination variable will receive the position of the item in the list (beginning with 0).

Examples:

b0 = 42

lookdown b0, (41, 42, 43, 44), b1

b0 = b0 + 2

lookdown b0, (41, 42, 43, 44), b2

In this example, b1 will receive the number 1 because 42 is in position 1 of the list and b2 will receive the number 3 because 44 is in position 3 of the list.

6

EXPERIMENTS

Basic Experiments

The basic experiments are all designed to run on the Schools Experimenter Board (AXE092). If you do not have a Schools Experimenter Board, then you can construct the circuit (see Fig. 6.1) on a prototype board and get exactly the same results. The Simple PIC board (AXE130) can also be used for many of the experiments, since its circuit is similar to the Schools Experimenter Board.

The Schools Experimenter Board, as its name suggests, was designed to be used as a teaching tool in schools. The board has space for a PICAXE-08M chip, programming socket and resistors, a light-dependent resistor, a press-button switch, three light-emitting diodes, and a piezo speaker. In addition, there is a switch that allows the light-dependent resistor, the press-button switch, and yellow and green LEDs to be disconnected from the PICAXE chip. The board also has provision for two header sockets that allow power and input/output pins to be connected to external devices.

The circuit for the Schools Experimenter Board is shown in Fig. 6.1; a list of materials is shown in Table 6.1.

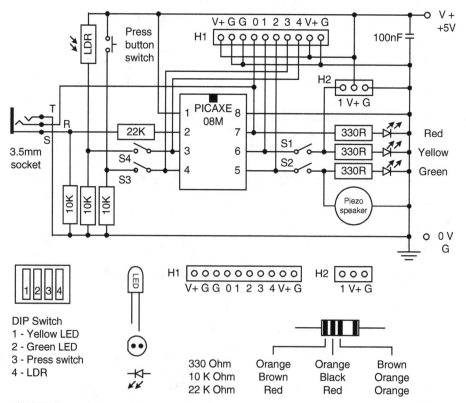

Figure 6.1 Circuit of the Schools Experimenter Board.

TABLE 6.1   PARTS LIST FOR SCHOOLS EXPERIMENTER BOARD

DESCRIPTION	PART NUMBER	QUANTITY
Printed circuit board	AXE092	1
3.5-mm stereo programming socket	CON039	1
Tactile push button switch		1
4-pole DIL switch		1
10-pin SIL socket		1
10-pin SIL header		1
100-nF (0.1-μF) bypass capacitor		1
Battery box with leads, 3 x AA or similar	BAT020	1
PICAXE-08M	AXE007M	1
Red LED 5 mm		1
Yellow LED 5 mm		1
Green LED 5 mm		1
22K resistor, $^1/_4$ W		1
10K resistor, $^1/_4$ W		3
330-Ω resistor, $^1/_4$ W		3
Light-dependent resistor 5 mm (LDR)	SEN002	1

Constructing the Schools Experimenter Board

You will need a fine-tipped temperature-controlled soldering iron, electrical solder, preferably lead-free, and wire cutters.

1 Start by identifying the components.

2 The circuit board has two sides: the copper and the component side with the component overlay printed on it. All components are mounted on the component side of the board with their leads passing through the holes in the board to the copper side, where they will be soldered.

3 Insert the resistors into the board and solder them in place. The leads are then cut off from the copper side of the board with wire cutters.

4 Insert the 8-pin IC socket and the tactile push-button switch and solder them into place. Note that the IC socket has a notch on one end and this should be oriented toward the notch symbol on the component overlay on the circuit board. The pins on the tactile switch form a rectangular pattern and the switch can only be fitted in two positions—either position is OK.

5 Insert the stereo socket, LDR, and LEDs and solder them into place. LED1 is red, LED2 is yellow, and LED3 is green. The numbers 0, 1, and 2 on the edge of the board correspond to the PICAXE pin numbers that the LEDs are connected to. Note that the LEDs are polarized and will not work if they are inserted the wrong way around. The LEDs have a flat side on the rim at the bottom of the body and this should be oriented to the flat, printed on the component overlay of the Schools Experimenter Board. The LDR is not polarized and can be mounted either way around. It may be advantageous to mount the LDR above the board surface so that it is not shaded by the stereo socket and switch and so that it can be bent at an angle. Cut the leads off from the copper side of the board with wire cutters once the LEDs and LDR have been soldered into place.

6 Insert the 100-nF capacitor and solder it; then, cut the leads off from the copper side of the board with wire cutters.

7 Insert the 4-way DIP switch and 10-way header socket and solder them into place. The 10-way SIL plug is used to plug into the socket and is not soldered to the board. (As a matter of individual preference, you can solder the SIL header to the board and use the socket to connect to the header.)

8 Connect the piezo speaker by placing both its leads through the inner hole in the board adjacent to the word "PIEZO." The hole in the corner of the board is there to allow a mounting screw to be attached. The ends of the piezo leads should be brought from underneath the board onto the component side of the board; they can be placed into the holes marked "B" and "R" for black and red, and then soldered. (Most piezo speakers have red and black leads. If yours has other colors that will not matter.)

9 Connect the battery leads by placing both leads through the inner hole in the board adjacent to the word "POWER." The hole in the corner of the board is there to allow a mounting screw to be attached. The ends of the battery box leads should be brought from underneath the board onto the component side of the board. They can then be placed into the holes marked "B" and "R" for the black and red leads, respectively. Take care with these leads. The circuit will not work if they are reversed.

10 At this point, it may be advantageous to perform steps 1 and 2 of "testing."

11 Using antistatic handling procedures, place the PICAXE chip into the IC socket. You may need to bend the pins inward slightly to make it fit. Note that the chip is polarized and the notch in the end must be aligned with the notch shown on the component overlay of the board (which will correspond with the notch in the 8-pin socket if it was installed correctly). When installed correctly, the notch in the PICAXE chip will point toward the programming socket. If the PICAXE chip is not installed the right way around it will not work and could be damaged.

Testing

1 Remove power from the board. Remove the PICAXE chip from the board using antistatic procedures. Reconnect power to the board.

2 Measure the voltage between pins 1 and 8 of the IC socket (see Fig. 6.2). It should be between 4.2 and 5.5 V for reliable operation; pin 1 must be +ve.

3 Turn all four DIP switches on.

4 Connect one end of a jumper wire to pin 1 of the IC socket and touch the other end to each of pins 7, 6, and 5 of the IC socket. You should see the red, yellow, and

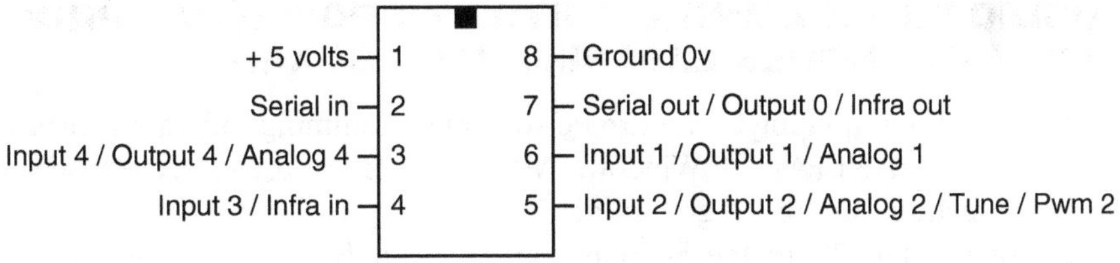

Figure 6.2 PICAXE-08M pin connections.

green LEDs light, respectively, as you do so. If they do not light, check that that the battery is connected and the DIP switches are turned on. Then, look for faults, such as the LEDs being connected the wrong way, a broken strip on the copper side of the board, or bad soldering. You may also hear the piezo speaker click when the green LED turns on.

5 Measure the voltage between pins 4 and 8 of the IC socket. It should be zero with the press-button switch released and equal to the battery voltage when the press-button switch is operated. If it's not, then check that the battery is connected and the DIP switches are turned on. Then, look for faults such as a broken strip on the copper side of the board, or bad soldering.

6 Measure the voltage between pins 3 and 8 of the IC socket. It should be zero when the LDR is dark and close to the battery voltage when the LDR is brightly lit. If not, first check that the battery is connected and the DIP switches are turned on. Then, look for faults such as a broken strip on the copper side of the board or bad soldering. Placing a finger on the LDR is a common way to make it dark, but be warned that the skin resistance of a finger can short-circuit some LDR's and give false readings.

IF IT DOESN'T WORK

1 Carefully check that all components are mounted in the correct place and with the correct polarity. The PICAXE chip, LEDs, and battery are polarized and will not work if they are mounted the wrong way. The resistors have different values and must be mounted in the correct place (see Fig. 6.1 for resistor codes).

2 Carefully check all the soldering for open circuits (broken/lifted copper or dry joints) and short circuits (solder has spilled onto adjacent tracks).

3 Ensure that there are fresh batteries inserted correctly in the battery box and that the battery box is connected and turned on.

4 Ensure that the PICAXE chip is correctly inserted into the IC socket with the notch facing the stereo connector. Check that none of the pins are broken off, bent up underneath the chip, or sticking out the side of the socket.

5 Follow the procedures under the heading "Testing."

6 Follow the procedures under the heading "If Program Downloading Fails."

In addition, the Schools Experimenter Board and all PICAXE chips can be programmed with the AXE028 USB programmer and over TCP/IP link.

USING PROGRAMMING EDITOR TO DOWNLOAD PROGRAMS TO THE SCHOOLS EXPERIMENTER BOARD

1 Install Programming Editor to a PC (other programming software such as Logicator for PIC or AXEPAD for Windows, Macintosh, or Linux can be used; however, these instructions are for Programming Editor).

2 Connect the PC to the Schools Experimenter Board by means of a USB programming cable, USB–serial adaptor and serial programming cable, or a serial programming cable, as shown in Fig. 6.3. Other methods of downloading programs to PICAXE chips are discussed elsewhere in this book.

3 Ensure that power is applied to the Schools Experimenter Board.

4 Start Programming Editor on the PC.

5 Select "View" from the Programming Editor menu and then "Options" to bring up the Options panel. On the "Mode" tab of the "Options" panel, select "08M" and "4 MHz." Optionally, check "Show options on startup" at the bottom of the "Options" panel.

6 Click on the "Serial port" tab of the "Options" panel and select the serial port that you are using. If you are using a USB port, it will show up as a serial port in this panel when the USB cable is plugged in and the driver for the USB cable has been installed. Note that early versions of Programming Editor recognized USB cables only at program start up; using the latest version of Programming Editor is recommended for all the experiments in this book.

7 Click the "OK" button at the bottom of the "Options" panel.

8 Open a code window by selecting "File" from the main menu and then selecting "New" and "New BASIC," or by pressing Ctrl+N.

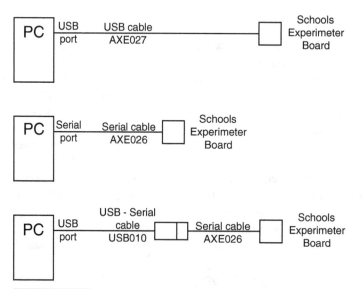

Figure 6.3 Downloading programs to the Schools Experimenter Board.

9 Type the following code into the code window.

```
begin:
    high 0
    pause 500
    low 0
    pause 500
    goto begin
```

10 Download the program to the Schools Experimenter Board by selecting "PICAXE" and "Program" from the Programming Editor menu, clicking the "Program" button in the Programming Editor taskbar, or by pressing F5 on the PC keyboard. You may see the red LED flashing during the download process—this is OK. When the program has finished downloading, there will be a short delay and the red LED will flash on and off at 1-second intervals. In some circumstances, it may be necessary to disconnect the programming cable from the Schools Experimenter board before the program will run.

IF PROGRAM DOWNLOADING FAILS

1 Check that a programming cable is connected between the PC and the Schools Experimenter Board as shown in Fig. 6.3.

2 Check that the Schools Experimenter Board has power applied.

3 Check that the Schools Experimenter Board has been correctly assembled.

4 Check that you have "08M" and "4 MHz" selected on the "Mode" tab of the "Options" panel in Programming Editor.

5 If using a serial port, check that the serial port in the Programming Editor "Options" panel is the same port that the programming cable is plugged into. If using USB, check that the USB cable drivers are installed on the PC. Check that the USB cable shows up as serial port in the Programming Editor "Options" panel. The USB cable drivers are available on the Programming Editor CD or online. Note that early versions of Programming Editor recognized USB cables only at program start up; using the latest version of Programming Editor is recommended for all the experiments in this book.

6 Try another programming cable.

7 Try programming the Schools Experimenter Board from another computer.

8 Try programming another Schools Experimenter Board.

CONNECTING EXTERNAL DEVICES

External devices can be connected to the Schools Experimenter Board by means of the 10-pin header socket (H1), which makes all of the PICAXE input/output pins available. Battery and ground pins are available on H1 and also on H2. By turning off the DIP switches, the yellow LED, green LED, LDR, and the press-button switch can be isolated from the PICAXE; the pins can then be used by an external devices via H1. There is no switch for the red LED; however, this output can still be used for external devices provided the output current does not exceed the limit of about 20 mA for an output pin.

The completed Schools Experimenter Board can be seen in Fig. 6.4. The PICAXE chip has not yet been inserted into its socket and the header 2 (H2) socket is not installed.

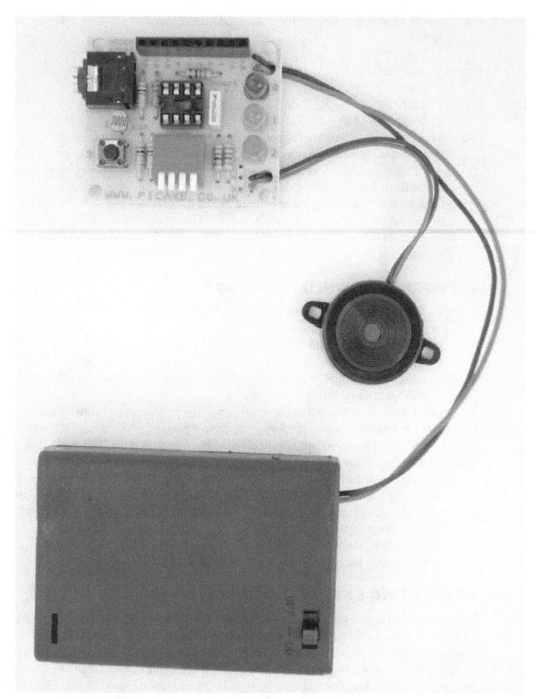

Figure 6.4 The completed Schools Experimenter Board.

FLASH THE RED LED

The following code will flash the red LED on the Schools Experimenter Board at 1-second intervals.

```
'Flash the red LED at 1-second intervals
#picaxe 08m
symbol redled = 0              'Define the pin number of the red LED
symbol delay = 500             'Define a delay of 500 (ms)
    do                         'Begin a "do ... loop" command
       high redled             'Turn the red LED on
       pause delay             'Wait for 500 ms with the red LED on
       low redled              'Turn the red LED off
       pause delay             'Wait for 500 ms with the red LED off
    loop                       'End a "do ... loop" command
```

Code analysis

Line	Command	Analysis
1	'Flash the red LED at 1-second intervals	This line is a comment because it begins with an inverted comma ('). It is used for documentation, in this case to give a title to the program.
2	#picaxe 08m	This line is a complier directive to set the mode to PICAXE-08M.
3	symbol redled = 0	Define a symbol called **redled** and set it equal to 0 (zero), which is the PICAXE pin that the red LED is connected to. You can now use the word **redled** in your program and the compiler will substitute 0 (zero) for it.
4	symbol delay = 500	Define a symbol called **delay** and set it equal to 500. You can now use the word **delay** in your program and the compiler will substitute 500 for it.
5	do	Define the beginning of a **do ... loop** command which ends in line 10. The **do ... loop** command will execute the commands between **do** and **loop** while a specific condition exists or until a specific condition occurs. In this program an ending condition is not specified so the loop will never exit.
6	high redled	Set pin 0 (zero) to the high state. The red LED is connected to pin 0 (leg 7) and will turn on when the pin is set to the high state. The number 0 (zero) is substituted for the word **redled**, because **redled** has been set to the value 0 (zero) in a **symbol** command.

7	pause delay	Pause program execution for 500 ms. The red LED will remain on for this period. The number 500 is substituted for the word **delay**, because **delay** has been set to the value 500 in a **symbol** command.
8	low redled	Set pin 0 (zero) to the low state. The red LED is connected to pin 0 (leg 7) and will turn off when the pin is set to the low state. The number 0 (zero) is substituted for the word **redled**, because **redled** has been set to the value 0 (zero) in a **symbol** command.
9	pause delay	Pause program execution for 500 ms. The red LED will remain off for this period. The number 500 is substituted for the word **delay**, because **delay** has been set to the value 500 in a **symbol** command.
10	Loop	End the **do ... loop** command, which began on line 5. It will cause the program to return to the command after the **do** command on the line 5, thus repeating the on-off sequence for the red LED. In this program an ending condition is not specified for the **do ... loop** command so the loop will never end.

FLASH THE YELLOW LED

The following code will flash the yellow LED on the Schools Experimenter Board at $3/4$-second intervals.

```
'Flash the yellow LED at 3/4-second intervals with different on and off periods.
#picaxe 08m

symbol yellowled = 1
symbol ondelay = 500
symbol offdelay = 250
    do
        high yellowled
        pause ondelay
        low yellowled
        pause offdelay
    loop
```

Code analysis

The code to flash the yellow LED works in that same way as the code to flash the red LED. The only differences are that pin1 (leg 6) is used instead of pin 0 (leg 7) and the on and off periods are different. Some blank lines have also been inserted into the code to enhance readability; the blank lines have no effect on the program.

FLASH THE RED AND GREEN LEDS ALTERNATELY

The following code will flash the red and green LEDs alternately on the Schools Experimenter Board at 1-second intervals.

```
'Flash the red and green LEDs alternately
#picaxe 08m

symbol redled = 0
symbol greenled = 2
symbol delay = 500
    do
        high redled
        low greenled
        pause delay
        low redled
        high greenled
        pause delay
    loop
```

The last three experiments contain enough information to give you the skills to make the LEDs flash in any combination; we now move on to something else.

Play Sounds through the Piezo Speaker

The **sound** command can play tones through the piezo speaker and the following code plays a selection of tones.

```
'Play sounds through the piezo speaker
#picaxe 08m
```

```
symbol lowtone = 50        'Define a value for the lowest tone
symbol hightone = 120      'Define a value for the highest tone
symbol period = 10         'Define a value for the period of the tone
symbol piezo = 2           'Define the pin that the piezo speaker is connected to
symbol tone = b0           'Define a variable to use for the loop counter and tone

    do

        for tone = lowtone to hightone

            sound piezo, (tone, period)

        next tone

        pause 500

    loop
```

It is normal for the green LED to flash while playing sounds through the piezo speaker.

Code analysis

Line	Command	Analysis
1	'Play sounds through the piezo speaker	This line is a comment because it begins with an inverted comma ('). It is used for documentation, in this case to give a title to the program.
2	#picaxe 08m	This line is a complier directive to set the mode to PICAXE-08M.
3		This is a blank line and is used to enhance readability of the code; it is ignored by the compiler.
4–8	symbol lowtone = 50 symbol hightone = 120 symbol period = 10 symbol piezo = 2 symbol tone = b0	Define symbols for the low tone, high tone, period of the tone, the PICAXE pin that is connected to the piezo speaker, and a memory variable to use as a counter for the **for ... next** command.
9		This is a blank line and is used to enhance readability of the code; it is ignored by the compiler.
10	do	Begin a **do ... loop** command, which ends on line 15. The **do ... loop** command will execute the commands between **do** and **loop** while a specific condition exists or until a specific condition occurs. In this program an ending condition is not specified so the loop will never end.

11	for tone = lowtone to hightone	Begin a **for ... next** command, which ends on line 13. The **for ... next** command will execute the commands between **for** and **next** while the loop counter is less than or equal to the end value (**hightone**). Symbol commands define **tone** as **b0, lowtone** as **50,** and **hightone** as **120** and, therefore, this loop will assign the value 50 to the loop counter b0 and increment the value of b0 by one for each iteration of the loop until b0 becomes greater than 120.
12	sound piezo, (tone, delay)	This command generates the sound. Symbol commands define pin 2 (leg 5) as the output pin, the value in the variable **b0** as the frequency of the tone, and a period of 10 ms for the tone.
13	next tone	End the **for ... next** command that began on line 11. Program execution will return to the **for** command in line 11.
14	pause 500	Pause program execution for $1/2$-s (500 ms).
15	loop	End the **do ... loop** command, which began on line 10. In this program an ending condition is not specified so the loop will never end thus repeating the series of tones.

You can also play tunes with the **play** command and you can make your own music with the **tune** command.

The following code uses the **play** command to play 2 of the 4 built-in tunes.

```
'Play Jingle Bells and Silent Night.
#picaxe 08M

symbol jingle_bells = 1
symbol silent_night = 2
symbol noleds = 0
symbol redled = 1

    play jingle_bells, noleds        'Play Jingle Bells
    pause 1000
    play silent_night, redled        'Play Silent Night and flash the red led
```

Digital Input

To present a digital input to the PICAXE, we must apply a voltage to an input pin that is within the range for digital inputs (0 to +0.8 V for logic low and +2 to +5 V for logic high). The Schools Experimenter Board and the Simple PIC Kit Board have a press-button switch connected to input 3 (leg 4) and a light-dependent resistor connected to input 4 (leg 3); these can both be used for digital input.

The push-button switch achieves logic low by connecting the PICAXE input pin to 0 (zero) V via a 10-K resistor when the switch is in the normal (not pressed) state. When the switch is pressed, the PICAXE input is connected directly to the positive supply through the switch contacts and this will present a logic high to the PICAXE input. The following program demonstrates the use of a switch for digital input by turning the red LED on when the switch is pressed.

TURN THE RED LED ON IF THE SWITCH IS PRESSED

'Digital input using a press-button switch	
#picaxe 08M	
symbol logichigh = 1	'Define the state for logic high
symbol redled = 0	'Define the pin number of the red LED
symbol switchin = pin3	'Define the pin number for the switch input
symbol switchpressed = logichigh	'Define switch-pressed state as "logic high"
do	'Begin a "do ... loop" command
if switchin = switchpressed then	'Begin an if command
	'and determine the state of the switch
high redled	'the switch is pressed, turn the red LED on
else	
low redled	'the switch is not pressed, turn the red LED off
endif	'End the "if" command
loop	'End the "do ... loop" command

Code analysis

Line	Command	Analysis
1	'Digital input using a press-button switch	This line is a comment, because it begins with an inverted comma ('). It is used for documentation, in this case to give a title to the program.
2	#picaxe 08m	This line is a complier directive to set the mode to PICAXE-08M.

4–7	symbol logichigh = 1 symbol redled = 0 symbol switchin = pin3 symbol switchpressed = logichigh	Define symbols for the logic high state, red LED output pin, switch input pin, and switch pressed state.
9	do	Begin a do . . . loop command, which ends on line 16. The do . . . loop command will execute the commands between do and loop while a specific condition exists or until a specific condition occurs. In this program an ending condition is not specified so the loop will never end.
10	if switchin = switchpressed then	Begin an if . . . then . . . else . . . endif command and specify the condition switchin = switchpressed. Symbol commands define switchin as the input pin that the switch is connected to and switchpressed as the state of the input pin when the switch is pressed. When the condition switchin = switchpressed is true, the command after the word then is executed.
11	'and determine the state of the switch	This line is a comment
12	high redled	Set pin 0 (zero) to the high state. The red LED is connected to pin 0 (leg 7) and will turn on when the pin is set to the high state. The number 0 (zero) is substituted for the word redled, because redled has been set to the value 0 (zero) in a symbol command.
13	else	This line is part of the if . . . then . . . else . . . endif command and, if the condition is not true, the command(s) following the else clause will be executed.
14	low redled	Set pin 0 (zero) to the low state. The red LED is connected to pin 0 (leg 7) and will turn off when the pin is set to the low state. The number 0 (zero) is substituted for the word redled, because redled has been set to the value 0 (zero) in a symbol command.
15	endif	End the if . . . then . . . else . . . endif command that began on line 10.
16	loop	End the do . . . loop command, which began on line 9. In this program an ending condition is not specified so the loop will never end.

TURN THE RED LED ON IF DARK

The LDR and 10-K resistor, connected between it and 0 (zero) V, form a voltage divider, and, when the resistance of the LDR is sufficiently low, the voltage at the junction of the LDR and 10-K resistor will fall into the range for logic high. Similarly, when the resistance of the LDR is sufficiently high, the voltage at the junction of the LDR and 10-K resistor will fall into the range for logic low. The following program demonstrates the use of an LDR for digital input by turning the red LED on when the LDR is dark.

```
'Digital input using an LDR
#picaxe 08M

symbol logiclow = 0          'Define the state for logic low
symbol redled = 0            'Define the pin number of the red LED
symbol ldrin = pin4          'Define the pin number for the LDR
symbol dark = logiclow       'Define the dark state as logic low

        do                   'Begin a "do...loop" command
           if ldrin = dark then   'Begin an "if" command
                                   'and determine the state of the LDR
               high redled       'If dark, turn the red LED on
               else
               low redled        'If light (not dark), turn the red LED off
           endif                'End the "if" command
        loop                 'End the "do...loop" command
```

Code analysis

Line	Command	Analysis
1	'Digital input using an LDR	This line is a comment, because it begins with an inverted comma ('). It is used for documentation, in this case to give a title to the program.
2	#picaxe 08m	This line is a complier directive to set the mode to PICAXE-08M.
4–7	symbol logiclow = 0 symbol redled = 0 symbol ldrin = pin4 symbol dark = logiclow	Define symbols for the logic low state, red LED output pin, LDR input pin, and dark state.
9	do	Begin a **do...loop** command, which ends on line 16. The **do...loop** command will execute the commands between **do** and **loop** while a specific condition exists or until a specific condition occurs. In this program, an ending condition is not specified so the loop will never end.

10	if ldrin = dark then	Begins an **if ... then ... else ... endif** command and specifies the condition **ldrin = dark.** Symbol commands define **ldrin** as the input pin that the LDR is connected to and **dark** as the state of the input pin when the LDR is in the dark state. When the condition **ldrin = dark** is true, the command after the word **then** is executed.
11	'and determine the state of the LDR	This line is a comment
12	high redled	Set pin 0 (zero) to the high state. The red LED is connected to pin 0 (leg 7) and will turn on when the pin is set to the high state. The number 0 (zero) is substituted for the word **redled,** because **redled** has been set to the value 0 (zero) in a **symbol** command.
13	else	This line is part of the **if ... then ... else ... endif** command and, if the condition is not true, the command(s) following the **else** clause will be executed.
14	low redled	Set pin 0 (zero) to the low state. The red LED is connected to pin 0 (leg 7) and will turn off when the pin is set to the low state. The number 0 (zero) is substituted for the word **redled,** because **redled** has been set to the value 0 (zero) in a **symbol** command.
15	endif	Specifies the end of the **if ... then ... else ... endif** command that began on line 10.
16	loop	End the **do ... loop** command, which began on line 9. In this program, an ending condition is not specified so the loop will never end.

Analog Input

Analog quantities are continuously varying quantities, such as sound or temperature, which may take any value between their limits. They differ from digital values, which have distinct states. The PICAXE has commands that can convert analog values to a number, which can be used to represent the analog quantity.

Turn the Red, Green, and Yellow LEDs On Depending on Light Level

The LDR is an analog device and will vary its resistance according to the level of light falling on it. For a small LDR (about 5-mm diameter), the resistance will vary from about

2,000 Ω in bright light to about 2,000,000 Ω in the dark. The LDR and 10-K resistor form a voltage divider and the voltage at the junction of the LDR and 10-K resistor can be measured at a PICAXE analog input. The following program demonstrates this by lighting different combinations of red, green, and yellow LEDs, depending on the light level.

```
'Analog input with LED display
#picaxe 08M

symbol minreading = 0                    'Define the analog values for dark to light
symbol verylowreading = 32
symbol lowreading = 50
symbol mediumreading = 85
symbol highreading = 128
symbol maxreading = 255

symbol redled = 0                        'Define the pin numbers for the LEDs
symbol yellowled = 1
symbol greenled = 2

symbol analogport = 4                    'Define the analog port for the LDR
symbol analogreading = b0                'Define a variable for the analog reading
    do
        readadc analogport, analogreading        'Read the analog value
        select analogreading                     'Turn on combinations of LEDs
                                                      depending 'on the analog reading.

            case minreading to verylowreading
                high redled
                low yellowled
                low greenled
            case verylowreading to lowreading
                high redled
                high yellowled
                low greenled
            case lowreading to mediumreading
                low redled
                high yellowled
                low greenled
            case mediumreading to highreading
                low redled
                high yellowled
                high greenled
            case highreading to maxreading
                low redled
                low yellowled
                high greenled
        endselect
    loop
```

Code analysis

Line	Command	Analysis
1	'Analog input with LED display	This line is a comment, because it begins with an inverted comma ('). It is used for documentation, in this case to give a title to the program.
2	#picaxe 08m	This line is a complier directive to set the mode to PICAXE-08M.
4–9	symbol minreading = 0 symbol verylowreading = 32 symbol lowreading = 50 symbol mediumreading = 85 symbol highreading = 128 symbol maxreading = 255	Define the values to be used for the different light level ranges. You can vary these values to suit yourself. The values should be in the range 0 to 255 (inclusive) and should be kept in ascending order.
11–13	symbol redled = 0 symbol yellowled = 1 symbol greenled = 2	Define the pin numbers for the LEDs.
15	symbol analogport = 4 symbol analogreading = b0	Define the analog port number and the variable to be used for the analog reading (in this case b0).
18	do	Begin a **do . . . loop** command, which ends on line 43. The **do . . . loop** command will execute the commands between **do** and **loop** while a specific condition exists or until a specific condition occurs. In this program, an ending condition is not specified so the loop will never end.
19	readadc analogport, analogreading	Reads the analog value of the voltage at pin 4 (leg 3) into the variable **analogreading.** Symbol commands define **analogport** as 4, and **analogreading** as b0.
20	select analogreading	Begin a **select . . . endselect** command that ends on line 42. The **select . . . endselect** command specifies a PICAXE variable and contains **case** clauses that allow different command(s) to be executed depending on the value of the variable. The **case** clauses may contain an individual value or range-of-values. This program uses the range-of-values feature to classify light levels into different ranges according to intensity.

22	case minreading to verylowreading	This line is a **case** clause within the **select ... endselect** command. It specifies the range of values between **minreading** (defined as 0 in a **symbol** command) and **verylowreading** (also defined in a **symbol** command). The commands on lines 23 through 25 will be executed if the value of **analogreading** falls within the range **minreading to verylowreading.**
23–25	high redled low yellowed low greenled	These commands will only be executed if the previous **case minreading to verylowreading** clause on line 22 was evaluated to true. The commands turn the red LED on and the yellow and green LEDs off to indicate that the light level is very low. Program flow then moves to the **endselect** command on line 42.
26	case verylowreading to lowreading	This line is a **case** clause within the **select ... endselect** command, it specifies the range of values between **verylowreading** (defined in a **symbol** command) and **lowreading** (also defined in a **symbol** command). The commands on lines 27 through 29 will be executed if the value of **analogreading** falls within the range **verylowreading to lowreading.**
27–29	high redled high yellowed low greenled	These commands will only be executed if the previous **case verylowreading to lowreading** clause on line 26 was evaluated to true. The commands turn the red and yellow LEDs on and the green LED off to indicate that the light level is low. Program flow then moves to the **endselect** command on line 42.
30–41		Lines 30 through 41 specify additional cases and commands that operate in the same way as the cases in lines 22 and 26.
42	endselect	End the **select ... endselect** command that began on line 20. It indicates that there are no more cases to be evaluated and program flow moves to the next command.
43	loop	End the **do ... loop** command, which began on line 18. In this program, an ending condition is not specified so the loop will never end.

Turn On the Red, Green, and Yellow LEDs Using a Potentiometer

Analog input can also be demonstrated by the use of a potentiometer as the following experiment demonstrates. It is essentially the same as the previous experiment except that the analog input is derived from a potentiometer rather than an LDR/resistor voltage divider.

For this experiment, you will need a 10-K potentiometer connected between +5 and 0 V (ground) with the wiper connected to PICAXE pin4 (leg 3). The LDR on the Schools Experimenter Board must be isolated from the PICAXE input by turning off the DIL switch number 4 (SW4). The potentiometer is connected to header H1 as shown in Fig. 6.5.

The code is exactly the same as the code for previous experiment titled "Analog input with LED display." The only difference between these experiments is that the analog input is supplied from a potentiometer instead of a voltage divider formed by a resistor and LDR.

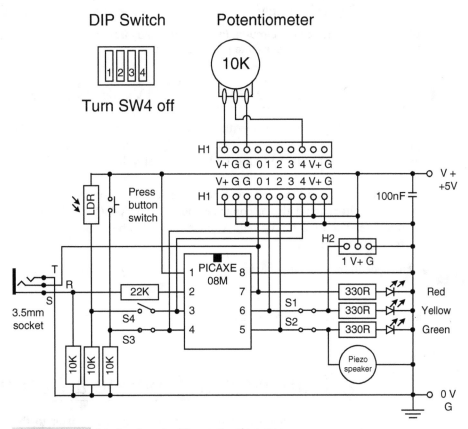

Figure 6.5 Analog input with a potentiometer.

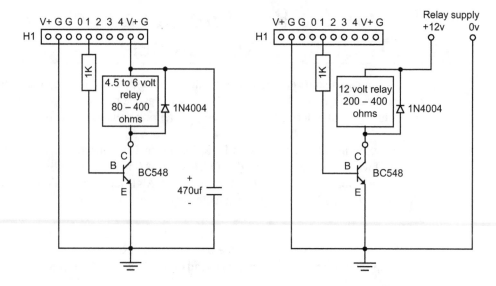

a. Driving a relay from the
PICAXE supply

b. Driving a relay from a
separate power supply

Figure 6.6 Relay driver circuits. (a) Driving a relay from the PICAXE supply;
(b) driving a relay from a separate power supply.

Using a Transistor to Drive a Relay

The circuits in Fig. 6.6 show how the transistors may be connected to the Schools
Experimenter Board in order to operate relays. Figure 6.6a shows a relay that runs from
the same supply as the PICAXE; Fig. 6.6b shows a relay that runs from a separate power
supply.

Note that if driving the relay from the same supply as the PICAXE (Fig. 6.6a), a
470-μF capacitor should also be connected between the V+ (5 V) and G (0 V) to avoid
any interference on the supply rail. A diode must also be connected across the relay
coil for either circuit.

The code to operate the relay is:

high 1

and the code to release the relay is:

low 1

Using a Transistor to Operate a Motor

The circuits in Fig. 6.7 shows a method of connecting a transistor to the Schools
Experimenter Board to drive a DC electric motor. Figure 6.7a shows a motor that runs

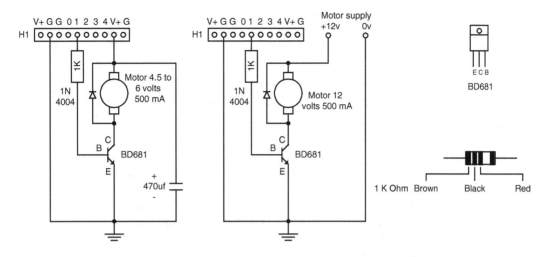

a. Motor driver using the PICAXE supply

b. Motor driver using a separate supply

Figure 6.7 Motor driver circuits. (a) Motor driver using the PICAXE supply; (b) motor driver using a separate power supply.

from the same supply as the PICAXE; Fig. 6.7b shows a motor that runs from a separate power supply. Note that if driving the motor from the same supply as the PICAXE, a 470-μF capacitor should also be connected between the V+ (5 V) and G (0 V) to avoid any interference on the supply rail. A diode must also be connected across the motor for either circuit.

Electric motors have a startup/stall current that is typically three to four times the motor's operating current and this must be allowed for when choosing a suitable transistor. The BD681 has sufficiently high current rating to drive a 500 mA motor and is used in these circuits.

The code to turn the motor on is

high 1

and the code to turn the motor off is

low 1

Controlling Motor Speed

By using the circuits in Fig. 6.7a or 6.7b with the following code, the speed of the motor can be varied from stopped to full speed. The speed is controlled by sending short variable-width pulses of full power to the motor. The pulses are sent so quickly that they are integrated together by the motor's inertia. The greater the ratio between the on and off periods, the more power is delivered to the motor and the faster it runs.

```
'Controlling motor speed
#picaxe 08M

symbol motor = 1                    'Define the motor port
symbol power = b0                   'Define a variable for the power level
symbol offperiod = b1               'Define a variable for the off period
symbol time = b2                    'Define a variable for a time period

    do                              'Begin a do ... loop command
      for power = 0 to 10           'Begin a for ... next loop for power level
        for time = 1 to 200         'Begin a for ... next loop for time period
          high motor                'Apply power to the motor
          pause power               'Wait for the on period to expire
          low motor                 'Remove power from the motor
          offperiod = 10 – power    'Calculate the off period
          pause offperiod           'Wait for the off period to expire
        next time
      next power
    pause 1000                      'Wait for 1 s and start the process again.

    loop
```

Code description

The code defines symbols for the motor port, power level variable, off period variable, and the time period. The program consists of three loops, which operate within each other. The first loop is a **do ... loop** command that does not have an exit condition and causes the program to repeat continuously. The next is a **for ... next** loop for the power level; the third loop is a **for ... next** loop for the time period.

The value of **power** variable determines the width of the pulses sent to the motor. There are 10 different values, which translate into 10 different power levels. A value of 0 gives 0% on time and 100% off time; a value of 10 will give 100% on time and 0% off time. The length of time the motor remains at each power level is determined by the value of the **time** variable. A value of 200 means that each power level will repeat 200 times, which gives sufficient time for the user to see the changes in motor speed.

Pulse-width modulation will be discussed further later in this book.

Controlling Motor Speed with a Potentiometer

A potentiometer can be used to control the speed of the motor by combining the motor driver circuits in Fig. 6.7 with the potentiometer circuit in Fig. 6.5 and the following code.

```
'Controlling motor speed with a potentiometer
#picaxe 08M

symbol motor = 1              'Define the motor port
symbol analogport = 4         'Define the analog port for the potentiometer
symbol analogreading = b0     'Define a variable for the analog reading
symbol offperiod = b1         'Define a variable for the off period

    do                                'Begin a continuous loop
      readadc analogport, analogreading   'Read the position of the potentiometer
      analogvalue = analogvalue / 20      'Reduce the range
      high motor                          'Apply power to the motor
      pause analogvalue                   'Wait for the on period to expire
      low motor                           'Remove power from the motor
      offperiod = 12 – analogvalue        'Calculate the off period
      pause off-period                    'Wait for the off period to expire
    loop
```

Code description

The motor on and off periods are controlled by the analog value, which represents the position of the potentiometer. The analog value will be an 8-bit number in the range 0 to 255; this range is reduced by dividing by 20 to produce 13 different values in the range 0 to 12. Each increment of the analog value will set a pulse on period of 1/13 of the time frame; the time frames are about 13 ms each.

This code can also be used to vary the brightness of a LED by using the circuit of Fig. 6.5, and to vary the brightness of an incandescent lamp by combining the circuits of Figs. 6.9 (see later) and 6.5.

Motor control and pulse-width modulation are discussed further later in this book.

Operating a Solenoid

The circuit of Fig. 6.8 shows a method of connecting a transistor to the Schools Experimenter Board to operate a DC solenoid.

The code to operate the solenoid is:

```
high 1
```

and the code to release the solenoid is:

```
low 1
```

It should be noted that many solenoids are designed for intermittent operation only and if left in the operated state for any length of time they may overheat and be permanently damaged.

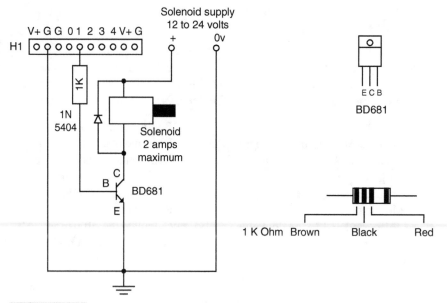

Figure 6.8 Transistor solenoid driver.

Operating an Incandescent Lamp

There is currently a global trend to phase out the use of incandescent lamps in favor of LED lamps, which in many applications can now provide the same levels of light at reduced power levels. However, there are still some incandescent lamps in use and this experiment is included for completeness.

The circuit in Fig. 6.9 shows a method of connecting a transistor driver to the Schools Experimenter Board to drive an incandescent lamp. Incandescent lamps have an in-rush current that is typically 8 to 10 times the lamps operating current. Therefore, a Darlington power transistor is used in this circuit to meet the increased current requirement.

The code to turn the lamp on is

high 1

and the code to turn the lamp off is

low 1

The brightness of the incandescent lamp can be controlled with a potentiometer by combining the circuits of Figs. 6.9 and 6.5 with the code for "Controlling motor speed with a potentiometer."

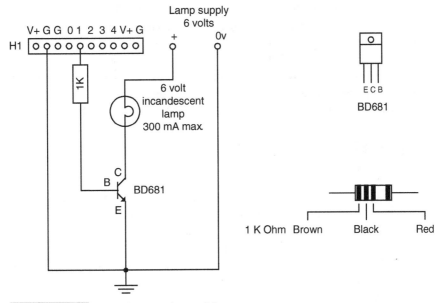

Figure 6.9 Incandescent lamp driver.

Water Detector

Figure 6.10 shows a method of connecting a transistor to the Schools Experimenter Board to detect the presence of a conducting liquid such as water. The output of the

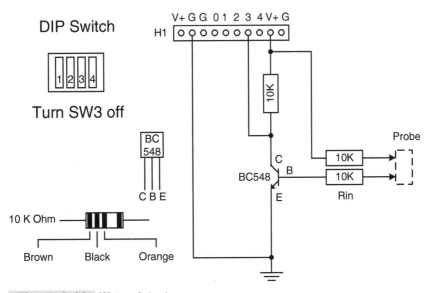

Figure 6.10 Water detector.

driver is connected to PICAXE input 3 (leg 4) so switch 3 on the DIP switch should be turned off to disconnect the push-button switch. The circuit as shown is very sensitive and can detect a drop of rainwater, which can have a resistance as high as 50,000 Ω. The circuit can be desensitized by increasing the value of Rin which has been split into two resistors in order to reduce the effect of electrolysis on the probe in the case where the container is earthed.

The code to turn on the red LED when water (or any conducting liquid) is present at the probe is:

```
'Water detector
#picaxe 08M

symbol probe = pin3      'Define the input pin for the water probe
symbol wet = 0           'Define the logic level for water present
symbol led = 0           'Define the output port for the red LED
do                       'Begin a loop
   if probe = wet then   'Begin an "if . . . endif" command and test for water
      high led           'Water is present so turn the LED on
   else
      low led            'Water is not present so turn the LED off
   endif
loop                     'No exit condition is specified so the code will loop
                             forever
```

7

ADVANCED EXPERIMENTS

Telephone Intercom	I2C Memory Expansion
Voltmeter	I2C I/O Expansion
1-Wire Serial Number	I2C Clock/Calendar
1-Wire Temperature	SPI Memory Expansion
Radio Frequency Identification	SPI I/O Expansion
Simple ASCII Terminal	UNI/O™ Memory Expansion

Bipolar Transistor Output Driver

The voltage and current available from a PICAXE output is not sufficient to directly drive many loads, such as relays, motors, solenoids, high-power light-emitting diodes (LEDs), and incandescent lamps. In order to drive these loads, a transistor driver can be added and NPN bipolar transistors or N-channel MOSFETs are both suitable and easy to connect to a PICAXE output.

The circuits of Fig. 7.1 show some practical circuits for transistor output drivers. With these circuits, a logic high on the PICAXE output pin will turn the load on; a logic low will turn the load off.

Interfacing Bipolar NPN Transistors

Bipolar transistors are current operated devices. The available load current from a bipolar transistor will be equal to the base current multiplied by the current gain (Hfe) of the transistor.

The data sheet for a BC548 shows that it has a maximum collector current of 500 mA and a minimum current gain of 110. Thus, the base current required to drive the maximum collector current is $500/110 = 4.5$ mA. The PICAXE outputs can supply about 20 mA and can, therefore, easily drive a BC548 to full output. Using a 1-K current-limiting resistor, as shown in Fig. 7.1a, will limit the current to around 4.5 mA when the PICAXE is operating from a 5-V supply; lower voltages will result in lower currents. This is marginal if the maximum collector current is needed and the 1-K resistor can be reduced as low as 330 Ω, if necessary.

For higher current loads, a transistor with higher current gain and current rating will be necessary. The data sheet for a BD681 shows that it has a maximum collector current of 4 A and a minimum current gain of 750. Thus, the base current required to

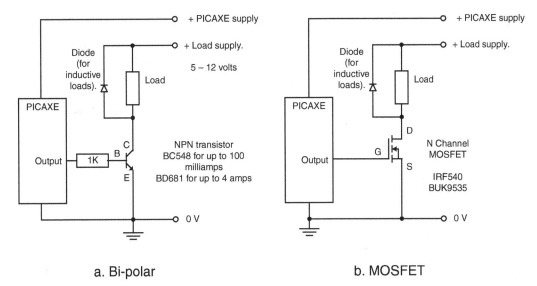

a. Bi-polar   b. MOSFET

Figure 7.1 Transistor output interface. (a) Bipolar; (b) MOSFET.

drive the maximum collector current is $4{,}000/750 = 5.3$ mA. The PICAXE outputs can supply about 20 mA and can, therefore, easily drive a BD681 to full output. Using a 1-K current-limiting resistor, as shown in Fig. 7.1, will limit the current to around 4.5 mA when the PICAXE is operating from a 5-V supply; lower voltages will result in lower currents. This is marginal if the maximum collector current is needed and the 1-K resistor can be reduced as low as 330 Ω, if necessary.

Note that it is always necessary to use a current-limiting device, usually a resistor, in series with the base of a bipolar transistor, that it is not always advisable to run a bipolar transistor at its maximum collector current and that power dissipation within the transistor must be taken into account.

Interfacing MOSFETs to a PICAXE

N-CHANNEL MOSFETS

MOSFETs typically need 10 to 15 V applied to the gate in order to achieve their minimum drain resistance. TTL devices can supply only 5 (best case) to 2.5 V (worst case). This means that a MOSFET requiring 10 V will not reach its minimum drain resistance when being driven directly from a PICAXE output. A MOSFET connected directly to a PICAXE output, as shown in Fig. 7.1b, can still be used for many applications, although the load current will be limited to about one-third of the maximum current available from the device.

If higher currents are needed, then a driver can be placed between the PICAXE and the MOSFET, bi-polar transistor drivers or high-voltage open-collector TTL gates can

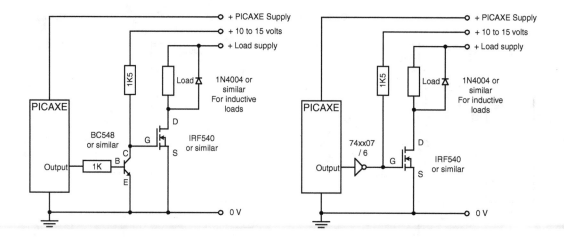

a Bi-polar driver   b TTL driver

Figure 7.2   MOSFET drivers. (a) Bipolar driver; (b) TTL driver.

perform this function, as shown in Fig. 7.2. Almost any general-purpose NPN transistor can be used and the 74LS07 hex buffer driver is a suitable open collector gate; it has six inverters in a package and the output driver is rated at 30 V.

The BUK9535 is designed to operate from TTL logic levels. It will provide full output current (34 A) with 5 V applied to the gate.

Transistor Input Driver

A transistor can also be connected as a driver to a PICAXE input to increase sensitivity, change voltage levels, and provide isolation. Figure 7.3 shows a simple method of connecting a transistor to a PICAXE input to increase sensitivity.

The sensitivity of the circuit can be varied by using different values of resistance for Rin. With the value shown (10 K), the circuit can accept up to +30 V at its input terminals and needs a minimum of around 0.5 V to produce a logic low at the PICAXE input terminal. The transistor inverts the input voltage and when there is a voltage at the input terminals (of the transistor), the circuit will present a logic low at the PICAXE input.

Interfacing TTL Chips

When operating from a supply voltage between 5 and 3.3 V, the PICAXE output logic levels are compatible with TTL levels and TTL devices can be connected to the PICAXE outputs with expectation that they will work.

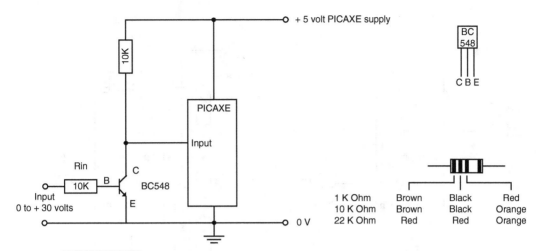

Figure 7.3   Transistor input driver.

However, the maximum voltage that can be applied to a PICAXE input is equal to the supply voltage plus 0.3 V. Thus, 5-V TTL levels should not be connected to a PICAXE running from a 3.3 V supply. In this case, a voltage divider can be used to reduce the voltage at the input pin, as shown in Fig. 7.4b. Digital interfacing and logic level conversion are discussed elsewhere in this book.

Integrated Driver Circuits

The ULN2803A is one of many integrated driver circuits designed to drive non-TTL loads (see Fig. 7.5). It is an array of eight inverters with TTL inputs, open-collector outputs, and back EMF suppression diodes built in. The outputs are rated at 50 V and 500 mA. The ULN2003 is a similar chip in a 14-pin package and contains 7 Darlington drivers.

H Bridge Motor Drivers L293, SN754410, L298

An H bridge consists of four switching devices, such as bipolar transistors or MOSFETs. H bridges are commonly used for driving DC and stepper motors, although their use is not limited to these applications.

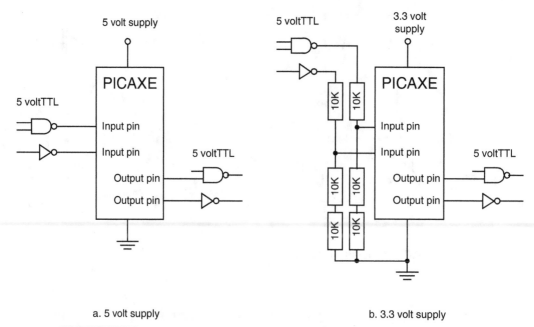

a. 5 volt supply   b. 3.3 volt supply

Figure 7.4 Interfacing TTL logic. (a) 5-V supply; (b) 3.3-V supply.

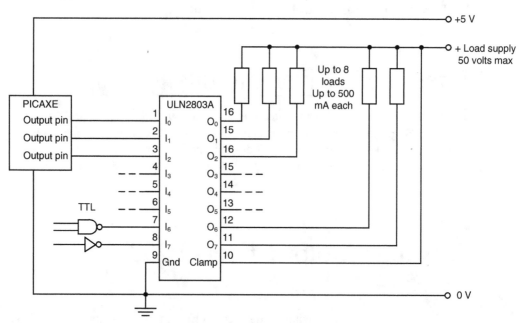

Figure 7.5 ULN2803 eight-way Darlington driver.

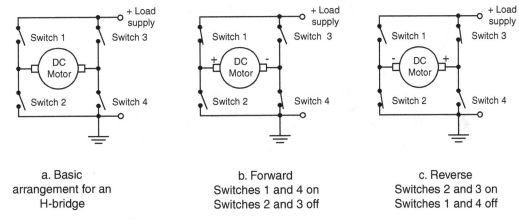

Figure 7.6 H-bridge motor driver. (a) Basic arrangement for an H bridge; (b) forward—switches 1 and 4 on and switches 2 and 3 off; (c) reverse—switches 2 and 3 on and switches 1 and 4 off.

Figure 7.6 illustrates the basic method of operation of H bridges and shows how it can be used to reverse the polarity of a DC motor. It is important to ensure that switches 1 and 2 and 3 and 4 are never turned on at the same time, since this will cause a short circuit between the supply rails. However, it is acceptable to turn on switches 1 and 3 or 2 and 4 at the same time; this principle can be used to provide dynamic braking for DC motors.

H-bridge circuits typically use MOSFETs as the switching elements and are often packaged in integrated circuits, such as the L293, SN754410, and L298, although they can also be constructed from discreet components.

Controlling Motor Speed and Direction with a Potentiometer

The following experiments show the use of the L293D and L298 integrated H-bridge circuits controlling the speed and direction of a DC motor using a potentiometer for speed control and pulse-width modulation to control the motor. Two circuits are shown in Fig. 7.7, one for the L293D or SN754410, and one for the L298.

The L293 and L298 each contain two H bridges, although only one has been used in the circuits shown. The L293D and SN754410 are pin compatible and either may be used. The L293D is capable of up to 600 mA output, while the SN754410 can provide output currents up to 1 A. The L298 has an output current capability of 2 or 4 A, if both drivers are connected in parallel.

The following code can be used with either of the circuits in Fig. 7.7 to vary the speed and control the direction of a DC motor using a potentiometer.

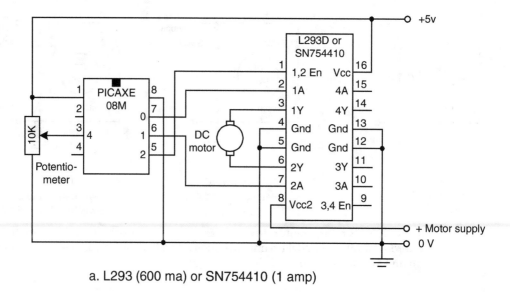

a. L293 (600 ma) or SN754410 (1 amp)

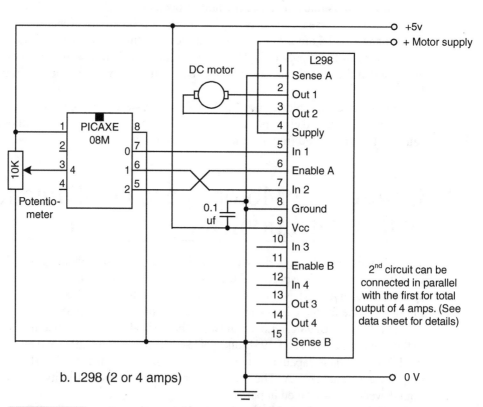

b. L298 (2 or 4 amps)

Figure 7.7 H-bridge motor controllers. (a) L293 (600 mA) or SN7544 (1 A); (b) L298 (2 or 4 A).

```
'Controlling motor speed and direction with a potentiometer
#picaxe 08m

symbol fwdmin = 492          'Analog reading for minimum forward speed
symbol revmin = 532          'Analog value for minimum reverse speed

symbol aport = 4             'Define port for potentiometer input
symbol pwmport = 2           'Define port for pulse stream output

symbol avalue = w0           'Define word for analog reading
symbol pwmonper = w1         'Define word for pwm dutycycle

    output 0                 'Set pin 0 as output
    output 1                 'Set pin 1 as output

do
    readadc10 aport, avalue                  'Read potentiometer position
    select avalue
        case 0 to fwdmin
            pins = $02                       'Set motor direction forward
            pwmonper = fwdmin - avalue * 21 / 10 MAX 1023 'Calculate on period
        case revmin to 1023
            pins = $01                       'Set motor direction reverse
            pwmonper = avalue - revmin * 21 / 10 MAX 1023 'Calculate on period
    endselect
    pwmout pwmport, 255, pwmonper
loop
```

The code keeps the motor at rest when the potentiometer is in the center position.
When the potentiometer is moved either left or right, the motor speed increases proportionally in the appropriate direction. The analog reading will be in the range 0 to
1,023 depending on the setting of the potentiometer. The halfway point is 512; readings
below halfway should turn the motor forward; readings above halfway should turn the
motor backward. In practice, a small window is left around the halfway point to make
it easier for the user to turn the potentiometer to the point where the motor will not
receive any power. Note that at low power levels the motor may not turn because of
limiting friction. The code can be modified to ensure that the minimum value for the
pulse on period is sufficient to start the motor.

The code commences by defining symbols for constants and variables and then
sets ports 0 and 1 to output. The code then enters a permanent loop that reads the
potentiometer setting from the analog port, determines which way to turn the motor,
sets the motor direction, calculates the duty cycle parameter (on period), and issues
the **pwmout** command. The code is the same for both the L293D (and SN754410) and
L298 circuits.

Stepper Motors

Stepper motors are DC motors that rotate in small steps of fixed amounts and, within the limits of their step angle, can be precisely positioned. They must be externally commutated because they have no commutator or brushes and, therefore, do not arc as they rotate. Stepper motors are suitable for use in places where precise positioning is required at modest speeds, such as positioning paper in printers. They may also be used in places where arcing would be hazardous.

Stepper motors fall broadly into two categories: unipolar and bipolar. Bipolar stepper motors have two windings, which must be reversed in polarity for rotation to occur. Unipolar stepper motors typically have four windings, which may be configured as four individual windings, two center-tapped windings, or two center-tapped windings with a common center tap, as shown in Fig. 7.8.

The operation, a stepper motor rotates a step at a time as pulses are applied to the windings in a specific sequence. For bipolar motors, the polarity of each winding must also be reversed, hence their name. In single-step mode, power can be removed and the motor will hold its position, provided no external force causes it to move. Leaving power connected provides greater holding torque, but increases power dissipation.

Stepper motors can generate more torque if two adjacent windings are energized at a time and they can also be made to rotate one-half step at a time by alternately energizing a single winding and then two adjacent windings. When used in these modes, power cannot be removed because the motor would fall back to a single-step position. Tables 7.1 and 7.2 show the stepping sequences for unipolar and bipolar stepper motors.

IDENTIFYING THE WINDINGS

The manufacturer's data sheet will provide a winding diagram for the motor, but if you've salvaged the motor you may not have access to the data sheet. In this case, a continuity checker can be used to identify the windings for bipolar stepper motors with four separate windings and unipolar stepper motors. For unipolar stepper motors with center-tapped windings, the resistance of the windings must be measured. The resistance from one end of the winding to the center tap will be one-half the resistance

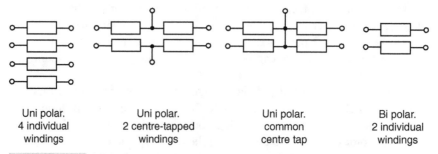

Uni polar.
4 individual
windings

Uni polar.
2 centre-tapped
windings

Uni polar.
common
centre tap

Bi polar.
2 individual
windings

Figure 7.8 Stepper motor winding arrangements. (a) Unipolar with four individual windings; (b) unipolar with 2 center-tapped windings; (c) unipolar with common center tap; (d) bipolar with two individual windings.

TABLE 7.1 STEPPING SEQUENCES FOR BIPOLAR STEPPING MOTORS			
UNIPOLAR STEPPER MOTORS. STEPPING SEQUENCES			
DRIVE TYPE	FORWARD SEQUENCE	REVERSE SEQUENCE	COMMENT
Single step			Least power dissipated, power can be removed after stepping
	0001	0001	
	0010	1000	
	0100	0100	
	1000	0010	
Single step high torque			Higher torque and power dissipation, power must be maintained to hold position
	0011	0011	
	0110	1001	
	1100	1100	
	1001	0110	
Half step			Step angle halved, torque varies, higher power dissipation, power must be maintained to hold position
	0001	0001	
	0011	1001	
	0010	1000	
	0110	1100	
	0100	0100	
	1100	0110	
	1000	0010	
	1001	0011	

of the entire winding. For unipolar stepper motors with four individual windings, the polarity of each winding may also need to be identified.

IDENTIFYING THE WINDING SEQUENCE

Unipolar and bipolar stepper motors will not rotate properly until the windings are connected in the correct sequence. It is relatively easy to determine the correct sequence by experiment; simply connect the windings in any sequence. If the motor does not rotate, then reverse connections 1 and 2. If the motor still does not rotate, then reverse connections 2 and 3. If the motor still does not rotate, then reverse connections 1 and 2 again.

CIRCUIT DESCRIPTION

Three circuits are shown; Fig. 7.9 is for a unipolar stepper motor using bipolar transistor drivers, Fig. 7.10 is for a unipolar stepper motor using an L293D motor driver, and

TABLE 7.2 STEPPING SEQUENCES FOR BIPOLAR STEPPING MOTORS

BIPOLAR STEPPER MOTORS. STEPPING SEQUENCES

DRIVE TYPE	FORWARD SEQUENCE	REVERSE SEQUENCE	COMMENT
Single step			Least power dissipated, power can be removed after stepping
	– – – +	– – – +	
	– – + –	+ – – –	
	– + – –	– + – –	
	+ – – –	– – + –	
Single step high torque			Higher torque and power dissipation, power must be maintained to hold position
	– – + +	– – + +	
	– + + –	+ – – +	
	+ + – –	+ + – –	
	+ – – +	– + + –	
Half step			Step angle halved, torque varies, higher power dissipation, power must be maintained to hold position
	– – – +	– – – +	
	– – + +	+ – – +	
	– – + –	+ – – –	
	– + + –	+ + – –	
	– + – –	– + – –	
	+ + – –	– + + –	
	+ – – –	– – + –	
	+ – – +	– – + +	

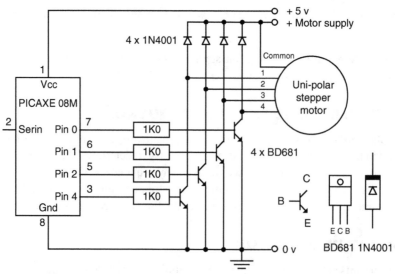

Figure 7.9 Unipolar stepper motor using bipolar transistor driver.

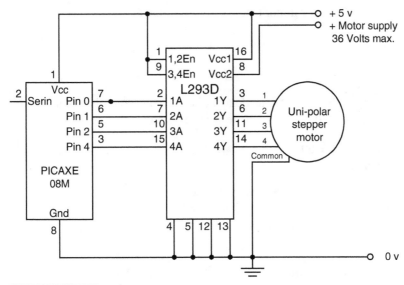

Figure 7.10   Unipolar stepper motor using L293 driver.

Fig. 7.11 is for a bipolar stepper motor also using an L293D motor driver IC. In each circuit, a PICAXE-08M is used to provide the four outputs for the stepper motor and software is used to provide the pulses.

The following code can be used with the circuits of Figs. 7.9, 7.10, and 7.11 to rotate the stepper motors through all the sequences given in Figs. 7.1 and 7.2.

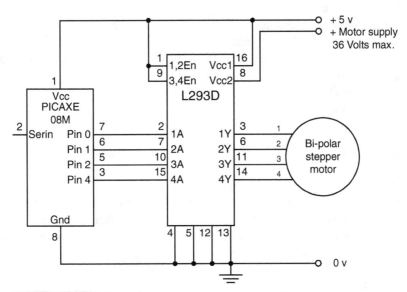

Figure 7.11   Bipolar stepper motor using L293 driver.

```
'PICAXE-08M stepper motor (same code for unipolar and bipolar)
# picaxe 08m
#freq m4

symbol singlestep = 4
symbol halfstep = 8
symbol delay = 100         'Time delay between steps, controls motor speed
symbol repeats = 20        'Step pattern repeats, causes the motor to turn continuously

symbol stepcount = b0
symbol counter = b1

    output 1               'Set pin 0 as output
    output 2               'Set pin 2 as output
    output 4               'Set pin 4 as output

do
    'Single step forward
    for counter = 1 to repeats            'Send the step pattern
        for stepcount = 1 to singlestep   'Generate the step pattern
            lookup stepcount, (0,$01,$02,$04,$10),   'Convert counter to step codes
                pins
            pause delay                   'Wait between steps
        next stepcount
    next counter

    pause 2000                            'Wait 2 s

    'Single step reverse
    for counter = 1 to repeats            'Send the step pattern
        for stepcount = 1 to singlestep   'Generate the step pattern
            lookup stepcount, (0,$01,$10,$04,$02),   'Convert counter to step codes
                pins
            pause delay
        next stepcount
    next counter

    pause 2000                            'Wait 2 s

    'Single step, high torque forward
    for counter = 1 to repeats            'Send the step pattern
        for stepcount = 1 to singlestep   'Generate the step pattern
            lookup stepcount, (0,$03,$06,$14,$11),   'Convert counter to step codes
                pins
            pause delay
        next stepcount
    next counter
```

```
    pause 2000                              'Wait 2 s

    'Single step, high torque reverse
    for counter = 1 to repeats              'Send the step pattern
        for stepcount = 1 to singlestep     "Generate the step pattern
            lookup stepcount, (0,$03,$11,$14,$06), pins  'Convert counter to step codes
            pause delay
        next stepcount
    next counter

    pause 2000                              "Wait 2 s

    'Half step forward
    for counter = 1 to repeats              "Send the step pattern
        for stepcount = 1 to halfstep       "Generate the step pattern
            lookup stepcount, (0,$01,$03,$02,$06,$04,$14,$10,$11), pins
            pause delay
        next stepcount
    next counter

    pause 2000                              "Wait 2 s

    'Half step reverse
    for counter = 1 to repeats              "Send the step pattern
        for stepcount = 1 to halfstep       "Generate the step pattern
            lookup stepcount, (0,$01,$11,$10,$14,$04,$06,$02,$03), pins
            pause delay
        next stepcount
    next counter

    pause 2000                              "Wait 2 s

loop                                        "Endless loop
```

Code analysis

The code is exactly the same for all three circuits. It commences by defining symbols for constants and variables, and then proceeds to turn the motors forward and backward using each of the stepping modes.

Servo Motors

Servo motors are a combination of an electric motor, a position-sensing device, such as a potentiometer, and a feedback circuit that controls the motor position. They are often used in radio-control applications for model cars, aircraft, and boats because of their ability to be positioned accurately and quickly.

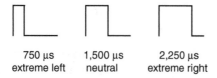

Figure 7.12 Servo pulse waveforms. (a) 750 μs, extreme left; (b) 1500 μs, neutral; (c) 2250 μs extreme right.

750 μs
extreme left

1,500 μs
neutral

2,250 μs
extreme right

A radio-control servo motor typically operates from a 4.8 or 6-V supply and can rotate through 180 to 210 degrees, although continuous rotation is used in some applications. Radio-control servos are controlled by a variable width pulse. Most servos are designed to place the output shaft in the center of its rotation when the pulse is 1,500 μs (1.5 ms) in duration. A 750-μs (0.75 ms) pulse will cause the shaft to move to one extreme and a 2,250-μs (2.25 ms) pulse will cause it to move to the other extreme. The pulses must be repeated at around 50 times per second for the servo to maintain its position. Servo pulses are illustrated in Fig. 7.12.

Servo motors typically have three leads: supply, pulse input, and a common or ground lead. In operation, the supply lead is connected to a positive supply and the input lead is connected to a source of pulses. A servo interfacing circuit is shown in Fig. 7.13. The simplicity of operation of servo motors makes them easy to control from a micro-controller, although to avoid interference, the PICAXE and the servo motor should be powered from separate supplies.

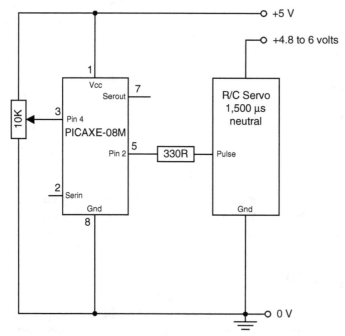

Figure 7.13 Servo interfacing circuit.

The following code uses a potentiometer to postion a radio-control servo that has a 1.5 ms neutral position.

```
'PICAXE-08M proportional servo control
# picaxe 08m
#freq m4

symbol servoport = 2
symbol analogport = 4
symbol analogvalue = w1

    servo servoport, 75          'Initialize the servo to center

do
    readadc analogport, analogvalue
    analogvalue = analogvalue * 10 / 17 + 75 'Divide by 1.7, add 75
    servopos servoport, analogvalue      'Position the servo
loop
```

Input and Output Expansion

Input/output port expansion is typically used where there are insufficient input/output ports available to connect a particular device, or where the device is remote from the processor and there is a need to reduce the number of connections between them. Shift registers and addressable I/O expander chips are two methods that can be used to expand the number of input/output ports. Figure 7.14 shows the use of shift registers for input and output expansion and addressable I/O expanders are illustrated in Fig. 7.17.

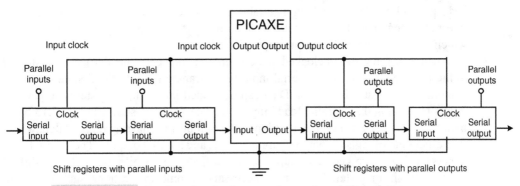

Figure 7.14 Input and output expansion using shift registers.

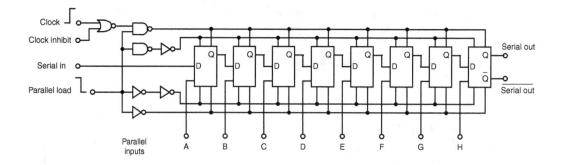

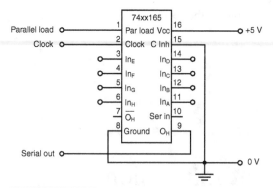

Figure 7.15 Internal arrangement of the 74xx165.

INPUT AND OUTPUT EXPANSION USING SHIFT REGISTERS

Shift registers are hardware devices that have a number of storage cells (called flip-flops) that can each store a single bit. Shift registers have serial input and output ports; there may also be a parallel input and/or output port. Parallel input ports allow all the cells to be loaded with an external value and parallel output ports allow the values in the cells to be transferred to an external device.

Figure 7.15 shows the internal arrangement for a 74xx165 8-bit shift register, which has a parallel input port, making it suitable for input expansion. Figure 7.16 shows the internal arrangement for a 74xx595 8-bit shift register that has a parallel output port, which makes it suitable for output expansion.

When a shift register is clocked, the contents of each cell (flip-flop) are transferred to the next cell, the bit on the serial input pin is transferred into the first cell, and the contents of the last cell are lost. Data can be loaded into a shift register with parallel inputs by pulsing the "parallel load" input pin. The 74xx165 shown in Fig. 7.15 has a parallel input port. Shift registers with parallel outputs may have an output latch, which is another kind of register that has parallel inputs and parallel outputs. The output latch, if present, is used to keep the output pins in a fixed state while data is being shifted through the shift register. This may be necessary in some applications, such as driving seven-segment LED displays. The 74xx595 shown in Fig. 7.16 has an output latch.

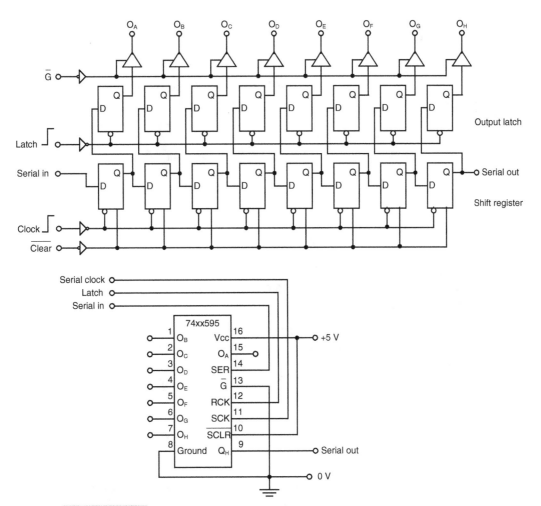

Figure 7.16 Internal arrangement of the 74xx595.

Driving seven-segment LED displays is an application where output expansion can be used to advantage. Seven-segment displays require seven pins per digit, or eight if the decimal point is used. It is quite practicable to use a PICAXE-08M to drive multidigit seven-segment displays with the addition of a shift register to expand the output capability.

Using shift registers for output expansion does have two disadvantages. A time delay occurs while the data is shifted into the register and the outputs are not true while the bits are being shifted through the cells. The time delay increases proportionally with the number of bits in the shift register. A practical limit will exist to the size of the shift register that can be used for an application. Using a latch can solve the "true output" disadvantage. The latch does not have to be a separate chip. Some shift registers, such as the 74xx595, have built-in latches (see Fig. 7.16).

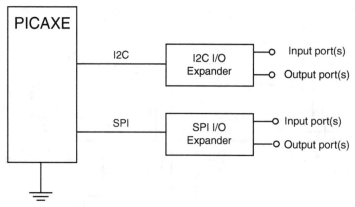

Figure 7.17 Input and output expansion using addressable I/O expander chips.

INPUT AND OUTPUT EXPANSION USING ADDRESSABLE I/O EXPANSION CHIPS

Addressable I/O expansion chips have a number of internal registers that can be directly addressed. Some of the registers are used for configuration and some are used for external port connections. (The use of addressable I/O expansion chips is demonstrated later in the I2C I/O expansion and SPI I/O expansion experiments in Figs. 7.50 and 7.53.)

Using Switches for Input

Switches are mechanical devices that use conductive contacts, usually made from metal, that are moved together to complete an electrical circuit. They are available in a wide variety of shapes, sizes, and configurations; it is difficult to describe them all in a single paragraph. Some common switch configurations are shown in Fig. 7.18. Double-throw switches sometimes have an off position (center off) between the two on positions. Switches with more than one pole are effectively multiple single-pole switches that are operated at the same time.

Switches are usually rated in terms of their contact configuration, the maximum voltage that can be safely applied to them, and the maximum current they can handle. Current-handling capacity can vary enormously from small tactile switches that are capable of only 10 mA to large toggle switches that can handle tens of amperes. When used to provide an input signal to a PICAXE, switches must be biased to either 0 V (ground) or +V to produce a TTL level signal, as shown in Fig. 7.19.

When a switch is opening or closing, the contacts tend to bounce and several circuit openings or closures may be made before the switch settles in a permanent state. For some applications, such as switching on a light or motor, bounce is not an issue, but the PICAXE operates quickly enough to be able to respond to bounces and it is often

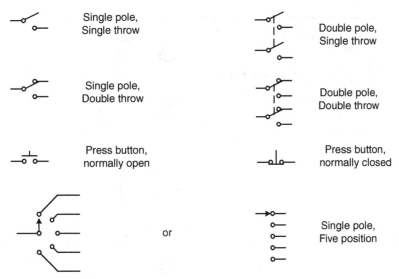

Figure 7.18 Common switch types.

necessary to debounce switches connected to PICAXE inputs. Switches typically settle into a permanent state within 20 ms.

Three methods of debouncing are commonly used; R/C networks, flip-flops, and software delays. Each method has advantages and disadvantages: hardware methods use extra components, software methods use extra code, R/C networks make use of the time delay that occurs when a capacitor is charged or discharged through a resistor and are reliable if appropriate values for R and C are chosen, flip-flop methods need no

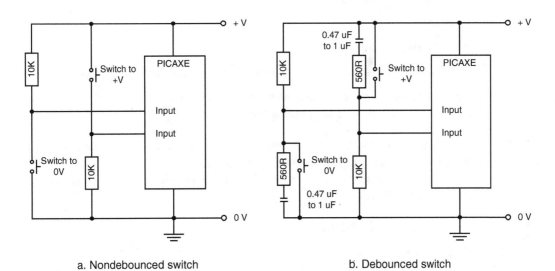

a. Nondebounced switch   b. Debounced switch

Figure 7.19 Switch debouncing with R/C network. (a) Nondebounced switch; (b) debounced switch.

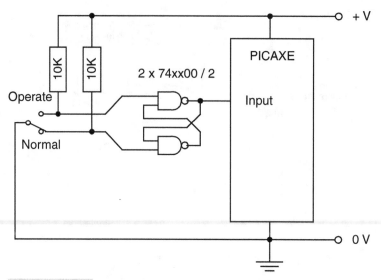

Figure 7.20 **Switch debouncing with a flip-flop.**

R/C network but require a double-throw switch and a flip-flop. Methods of debouncing switches using hardware are shown in Figs. 7.19 and 7.20. The 560-Ω resistor in series with the capacitor limits the current through the switch when it is closed and should not be omitted.

Software debouncing typically uses code to introduce a delay in order to allow the switch to stabilize; the following pseudocode shows a method of doing this:

Read the switch state and store it

Wait 20 ms

Read the switch state

If the switch state is the same as the stored state, then the switch is stable

If the switch state is not the same as the previous state, then the switch is not stable. Repeat the entire process until the switch is stable.

When connected to PICAXE inputs, switches are effectively digital input devices, as discussed in Chapter 3, Input and Output Techniques.

Suitable code to read the state of a switch connected to pin1 and 0 V (ground) as per Figs. 7.19 and 7.20 is:

if pin1 = 0 then switchdown

if pin1 = 1 then switchup

The switchdown and switchup states are reversed if the switch is connected to the +V supply.

Relays

Relays are electrically operated switches; some are mechanical and others are electronic (solid state). Mechanical relays use a solenoid to operate switch contacts and typically take between 2 and 20 ms to operate and release. Their contacts will bounce and the solenoid coil produces a back EMF when power is removed. Operating voltages of 5, 6, 12, 24, and 48 are common; other voltages are possible. Relays can be designed to operate from DC or AC and can be either monostable or bistable.

Monostable relays are the most common type of relay; they are mechanical in operation and have a single solenoid coil that operates switch contacts. Typical contact configurations are SPST, SPDT, DPST, and DPDT, although other contact configurations are possible. Monostable relays are in the normal, or at-rest, position when the solenoid coil is not energized and move to the operate position when the coil is energized. Power must be applied continuously to the coil to keep the relay in the operate state.

Bistable relays are often found in battery-operated equipment, because they require very little power. They are mechanical in operation and have two coils, one to place them in the operate state and the other to place them in the release state. There is usually a small biasing magnet to retain the current state when power is removed. They require only short pulses of current through the coils to operate or release. Typical contact configurations are SPST, SPDT, DPST, and DPDT, although other contact configurations are possible.

Solid-state relays use electronic switching and have no moving parts. They typically have 4,000 V, or more, of isolation between the input and output terminals. This makes them suitable for switching high voltages, such the AC power mains (solid-state relays are available in AC and DC versions). The input of a solid-state relay typically operates from any DC voltage between 3 and 32 V and is TTL-level compatible. Solid-state relays are silent in operation, do not generate back EMF, and operate within one-half an AC cycle with no contact bounce. Typical contact configuration is SPST; if other configurations are required, then several relays can be combined to increase the number of poles or throws.

Figure 7.21 shows practical circuits for operating monostable, bistable, and solid-state relays from a PICAXE. Note that the solid-state relay can be connected directly to a PICAXE output and does not require an external power supply.

The code to operate and release the relays is:

```
'Relay driver
#picaxe 08m


    high 0          'Operate monostable relay
    pause 500       'Keep it operated for 1/2 s
    low 0           'Release monostable relay
```

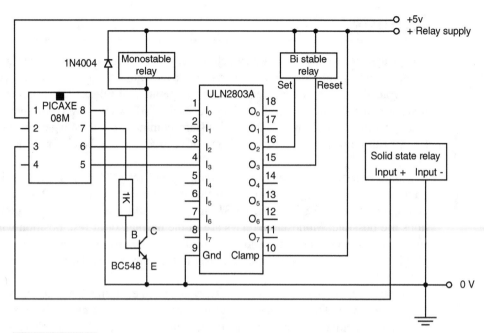

Figure 7.21 Relay driver circuits.

high 1	'Set bistable relay
pause 20	'Wait 20 ms to ensure relay is set
low 1	'Remove power from the "set" coil
pause 500	'Keep it "set" for half a second
high 2	'Reset bistable relay
pause 20	'Wait for 20 ms to ensure relay is reset
low 2	'Remove power from the "reset" coil
high 4	'Operate solid-state relay
pause 500	'Keep it operated for $^1/_2$ s
low 4	'Release solid-state relay

Wireless Links

A wireless link, as the name suggests, is a link between two or more devices that does not require wires. Wireless links may be implemented with light beams, usually infrared, or radio.

The PICAXE supports the Sony Infrared Control protocol (SIRC) for infrared links. Radio links are typically implemented using external transmitter/receiver modules and many radio modules that operate in the 433 MHz, 915 MHz, or 2.4 GHz bands are suitable for use with PICAXE chips. Infrared links typically have ranges up to 30 m

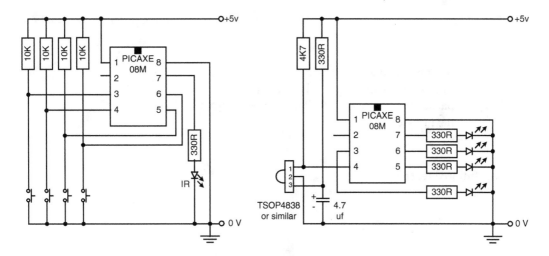

a. Infrared transmitter   b. Infrared receiver

Figure 7.22 Infrared wireless link. (a) Infrared transmitter; (b) infrared receiver.

and radio modules typically have ranges from several meters to several kilometers, depending on the transmitter power, receiver sensitivity, aerial system, and transmission path.

INFRARED

Figure 7.22 shows a practical circuit for an infrared wireless link. With the components shown, the range is about 30 cm (12 in); this can be increased by providing more current to the IR LED. A transistor driver can be used for this purpose.

The codes for the infrared transmitter and receiver are:

```
'Infra-red transmitter
#picaxe 08m

symbol down = 0            'Define logic level for switch pressed
symbol switch1 = pin1      'Define port for switch 1

symbol irbyte = b1         'Define a variable for a counter

    'Hold switch 1 down at power on to enter test loop
    if switch1 = down then testloop

    do                          'Main loop
        irbyte = pins / 2 xor $0F   'Align in low-order four bits and invert switch states
        infraout 3, irbyte          'Transmit a byte
    loop
```

```
testloop:                      'Test loop cycles through all the possible combinations

  do
      for irbyte = 0 to 15
      infraout 1, irbyte       'Transmit a byte
      pause 50                 'Pause between bytes
      next irbyte
  loop

'Infrared receiver
#picaxe 08m

symbol infradata = b0

  output 1, 2, 4                'Define pins 1, 2, and 4 as outputs

  do
      infrain2
      infradata = infra * 2 and $10 | infra and $17      'Move bit 3 to bit 4 for output
      pins = infradata
  loop
```

Code description

The transmitter has four switches and the state of these switches is transmitted as bits in a byte. The receiver has four LEDs, which are used to display the state of the switches on the transmitter. There is also a facility to test the operation of the transmitter and receiver by cycling through all possible combinations of the switch states. To enter the test state, hold down switch 1 when the transmitter is turned on; to exit the test state, turn the transmitter off and on again.

The transmitter code starts by defining the state for 'switch pressed,' the port for switch 1, and a variable for the state of the switches and counter. Switch 1 is then tested and if it is pressed the code enters the test loop.

The main loop reads the **pins** variable, which will contain the state of the switches in bits 4 to 1. For convenience, these are shifted into bits 3 to 0 by shifting right (dividing by 2), which are then inverted so that a switch that is pressed is represented by a 1 in the corresponding bit position. The resulting byte will have zeros in the four high-order bits and the state of the switches in the four low-order bits. The test loop uses a counter to generate all possible combinations of the four low-order bits.

The receiver code sets pins 1, 2, and 4 as outputs. It then enters an endless loop that reads a byte, moves the bit in position 3 to position 4, and sets the state of the output pins and, thus, light the corresponding LEDs. Note that bit 3 is moved to the bit 4 position for output because bit 3 of the **pins** variable is not implemented for output.

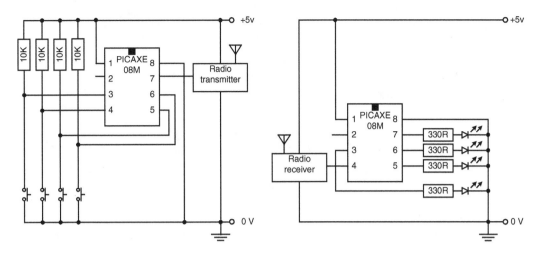

a. Radio transmitter

b. Radio receiver

Figure 7.23 Radio wireless link. (a) Radio transmitter; (b) radio receiver.

RADIO

Radio modules generally fall into two categories: those that transmit bits and those that transmit bytes. The modules that transmit bits tend to use a bit-slice detector in the receiver to reconstruct the data stream. Bit-slice detectors require an equal number of 1s and 0s in the bit stream when averaged over a short period of time, typically around 4 ms. This means that they are generally not suitable for transmitting ASCII directly, but require the use of Manchester codes or synchronizing characters. It is possible to achieve reasonably reliable results when using receivers with bit-slice detectors by transmitting frequent synchronizing characters that have the same number of one bits and zero bits such as hex. AA or hex. 55.

The circuit of Fig. 7.23 is suitable for radio transmitters that transmit and receive bits (bit-slice detector) or bytes.

The following transmitter and receiver code can be used when using radio devices with bit-slice detectors in the receivers. Note that the receiver may desynchronize from time to time with such a simple form of synchronization and improved synchronization or CRCs should be used in practice to ensure data integrity. This code can also be used with byte transmitters and receivers, although the bit synchronization characters are superfluous.

```
'Radio control - Bit-slice devices transmit
#picaxe 08m

symbol down = 0              'Define logic level for switch pressed
symbol switch1 = pin1        'Define port for switch 1
```

```
symbol serdata = b0              'Define a variable for the wireless data
symbol counter = b1              'Define a variable for a counter


     'Hold switch 1 down at power on (reset) to enter test loop
     if switch1 = down then testloop


do                               'Main loop
    serdata = pins / 2 xor $AA   'Read the state of the switches, shift right
                                 'to align in order four bits, add synchronizing
                                 'bits to high nibble, invert bits 3 and 1 to
                                 'approximate equal 1's and 0's.
    serout 0, T2400, (serdata)   'Transmit a byte
    pause 2                      'Allow receiver time to process but not to desynch

loop


testloop:                        'Test loop cycles through all the possible combinations
    do
        for counter = 0 to 15
            serdata = counter xor $AA 'Add some synchronizing bits and invert
                                 'bits 3 and 1 to approximate equal 1's and 0's
        serout 0, T2400, (serdata)    'Transmit a byte
        pause 10                 'Allow receiver time to process but not to desynch
        next counter
    loop


'Radio control - Bit-slice devices receive
#picaxe 08m


symbol serdata = b0              'Define pins 1, 2, and 4 as outputs

    output 1
    output 2
    output 4


do
    serin 3, T2400, serdata
    serdata = serdata xor $AA               'Remove sychronizing bits from high nibble
                                            'and invert bits 3 and 1 to restore their
                                            'state (which was inverted at the transmitter)
    serdata = serdata * 2 & $10 | serdata & $17      'Move bit 3 to bit 4 for output
    pins = serdata                          'Display the switch state on the LED's
loop
```

Code description

The transmitter has four switches and the state of these switches is transmitted in a byte, which has some synchronizing bits added and other bits inverted to approximate an equal number of 1s and 0s in the bit stream. The receiver has four LEDs which are used to display the state of the switches on the transmitter. There is also a facility to test the operation of the transmitter and receiver by cycling through all possible combinations of the switch states. To enter the test state, hold down switch 1 when the transmitter is turned on; to exit the test state, turn the transmitter off and on again.

The transmitter code starts by defining the state for 'switch pressed,' the port for switch 1, and a variable for the state of the switches and counter. Switch 1 is then tested and if it is pressed, the code enters the test loop.

The main loop reads the **pins** variable, which will contain the state of the switches in bits 4 to 1. For convenience, these are shifted into bits 3 to 0 by shifting right (dividing by 2). The resulting byte will have zero's in the four high-order bits and the state of the switches in the four low-order bits. It is then exclusive or'ed with hexadecimal AA (binary 10101010), which will leave hexadecimal A in bits 7 to 4 and invert bits 3 and 1. This process will construct a byte that has approximately the same number of 1s and 0s. The byte is then transmitted. The test loop does the same thing, except that it uses a counter to generate all possible combinations of four switches.

The receiver code starts by defining a variable for the received data and then sets pins 1, 2, and 4 as outputs. It then enters an endless loop that reads a byte, exclusive or's it with hexadecimal AA to remove the highorder hexadecimal A the transmitter inserted, and inverts bits 3 and 1 to their original state. The bit in position 3 is moved to position 4 and the byte is then used to set the state of the output pins and thus light the corresponding LEDs. Note that the LED will not be lit when the corresponding switch is pressed, because the switches present a logic high at the PICAXE inputs when they are not pressed. Note also that bit 3 is moved to the bit 4 position for output, because bit 3 of the **pins** variable is not implemented for output.

The circuits of Fig. 7.23 can also be used with byte radio modules, although the code can be simplified because there is no need to transmit equal numbers of ones and zeros and, therefore, the synchronizing characters are not necessary.

Light-Emitting Diodes

Small LEDs typically require around 1.2 to 3.4 V, depending on color, at currents of 5 to 20 mA. This means that they can be driven from PICAXE output pin, via a current-limiting resistor, without the need for an interface circuit. Some LEDs, such as high-brightness LEDs or larger seven-segment displays consisting of several LEDs connected in series or parallel, may require higher voltages or currents. Such devices require an interface circuit to supply the extra voltage and/or current needed.

LEDs may be single or several devices in a single package with a common lead; some possible arrangements are shown in Fig. 7.24. The circuits in Fig. 7.25 show several practical methods of connecting LEDs to a PICAXE.

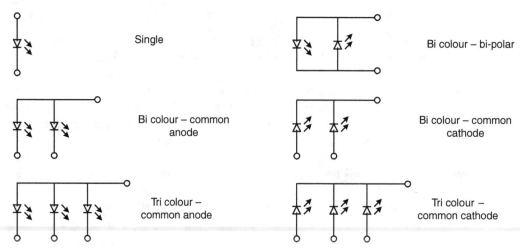

Figure 7.24 LED arrangements.

Seven-Segment LED Displays

Numeric seven-segment LED displays consist of eight LEDs, one per segment and one for the decimal point, with either a common anode or common cathode connection. The smaller displays use 20-mA LEDs, which makes them suitable for connection directly to a PICAXE port or TTL output. The larger displays employ multiple LEDs per segment which may be series or parallel connected.

The segments of a seven-segment display are designated by the letters "A" to "F," and the arrangement and segment patterns are shown in Figs. 7.26 and 7.27.

Figure 7.28 shows a means of connecting a seven-segment LED display directly to a PICAXE-18X. Figures 7.29 and 7.30 show a means of connecting seven-segment displays indirectly by using external shift registers.

The code to display the hexadecimal numbers 0 to F is :

```
'Display hexadecimal numbers 0 to F on seven-segment display using a parallel interface
#picaxe 18x

symbol outbyte = b3
symbol counter = b4

    do
    for counter = 0 to 15
        lookup counter, ($BE, $82, $DC, $D6, $E2, $76, $7E, $92, $FE, $F2,
          $FA, $6E, $3C, $DE, $7C, $78), outbyte
        pins = outbyte
        pause 500
    next counter
    loop
```

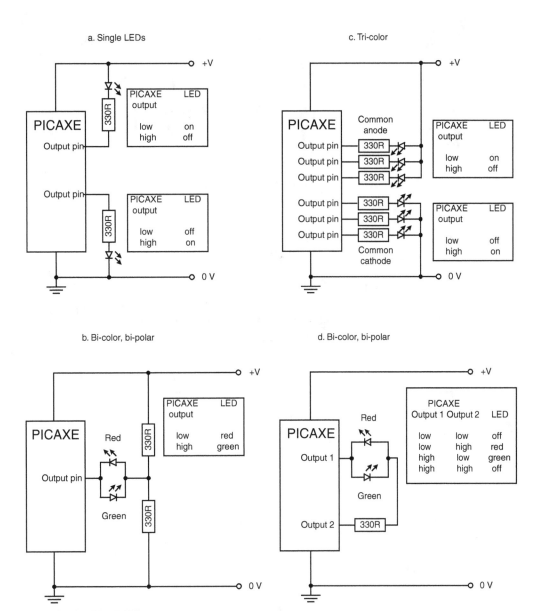

Figure 7.25 LED circuits. (a) Single LEDs; (b) bicolor, bipolar; (c) tricolor; (d) bicolor, bipolar.

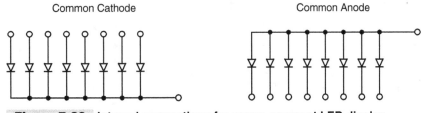

Figure 7.26 Internal connections for seven-segment LED display.

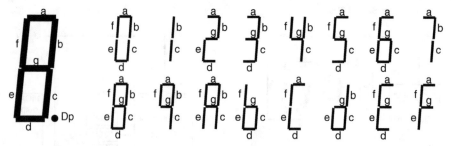

Figure 7.27 Segment arrangements for seven-segment displays.

The code uses a continuous loop to generate the hexadecimal numbers 0 to F (decimal 0 to 15). A lookup command is used to convert the numbers into segment codes that are written to the output pins.

The segment codes can be derived from Table 7.3. These segment codes are based on the connections in Fig. 7.27 between the display and PICAXE. It is both possible and practical to use different connections and, in that case, the segment codes will need to be recalculated for the new connections.

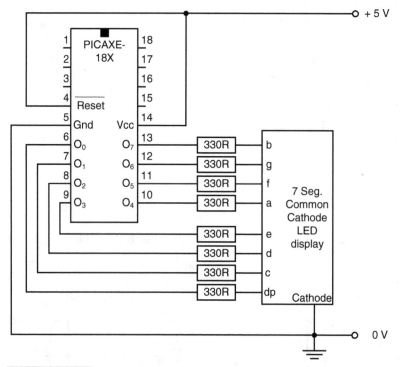

Figure 7.28 Connecting a seven-segment display directly to a PICAXE.

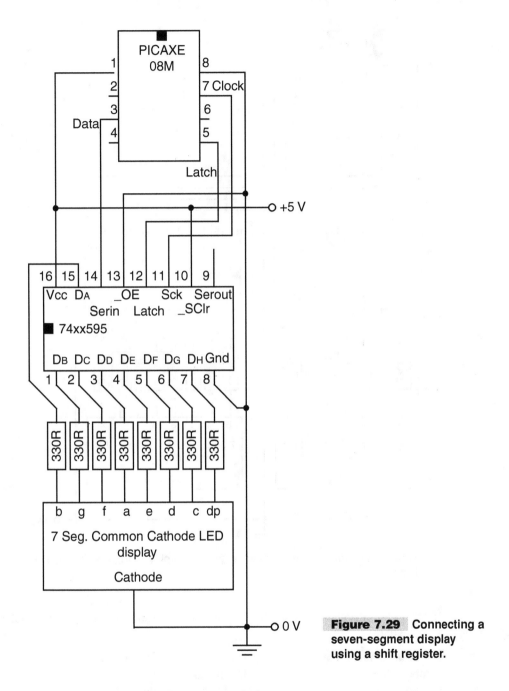

Figure 7.29 Connecting a seven-segment display using a shift register.

Figure 7.29 shows a means of connecting a seven-segment LED display to a PICAXE-08M using a shift register to expand the number of outputs. Several digits can be displayed by daisy-chaining shift registers. Figure 7.30 demonstrates this with a two-digit display. Note the connection between the 'serout' port of the first shift register and the 'serin' port of the second shift register.

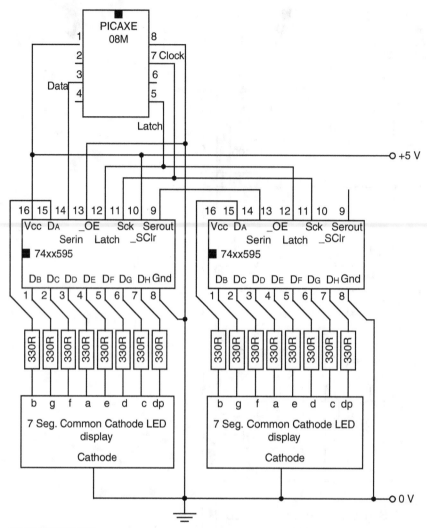

Figure 7.30 Connecting a seven-segment display using a shift register with two digits.

'Display hexadecimal numbers 0 to F on 7 segment display using a serial interface
#picaxe 08m

symbol dataout = 4   'Define port for data stream
symbol clk = 0   'Define port for clock
symbol latchout = 2   'Define port for latch

symbol bitcounter = b1   'Define variable names
symbol outbit = b2
symbol outbyte = b3
symbol counter = b4

TABLE 7.3   CODES FOR A SEVEN-SEGMENT DISPLAY

SEGMENT	B	G	F	A	E	D	C	dp	
Bit	7	6	5	4	3	2	1	0	
Digit									HEX.
0	1	0	1	1	1	1	1	0	BE
1	1	0	0	0	0	0	1	0	82
2	1	1	0	1	1	1	0	0	DC
3	1	1	0	1	0	1	1	0	D6
4	1	1	1	0	0	0	1	0	E2
5	0	1	1	1	0	1	1	0	76
6	0	1	1	1	1	1	1	0	7E
7	1	0	0	1	0	0	1	0	92
8	1	1	1	1	1	1	1	0	FE
9	1	1	1	1	0	0	1	0	F2
A (10)	1	1	1	1	1	0	1	0	FA
B (11)	0	1	1	0	1	1	1	0	6E
C (12)	0	0	1	1	1	1	0	0	3C
D (13)	1	1	0	1	1	1	1	0	DE
E (14)	0	1	1	1	1	1	0	0	7C
F (15)	0	1	1	1	1	0	0	0	78
Dec. pt.	0	0	0	0	0	0	0	1	01

```
do
    for counter = 0 to $F          'Count to hexadecimal F (decimal 15)
        gosub shift595             'Convert to segment codes and shift into 74xx595
        pulsout latchout, 1        'Latch the data
        pause 500
    next counter
loop
```

'Subroutine to convert a digit to segment codes and shift them into a 74xx595,
'the digit to be sent is in 'counter'

```
shift595:
    lookup counter, ($BE, $82, $DC, $D6, $E2, $76, $7E, $92, $FE, $F2,$FA, $6E,
        $3C, $DE,$7C, $78), outbyte

    for bitcounter = 0 to 7        'Count to 8
        outbit = outbyte & 1       'Isolate the bit to be sent
        if outbit = 1 then         'Test the bit, if it's a 1 then
            high dataout           'set the data out port high
        else                       'otherwise
            low dataout            'set the data out port low.
        endif
```

```
        pulsout clk, 1              'Clock the shift register
          outbyte = outbyte / 2    'Shift right to position the next bit for output
        next bitcounter
        return
```

The code uses a continuous loop to generate the hexadecimal numbers 0 to F (decimal 0 to 15), the digit to be sent is placed in the variable 'outdigit' and the subroutine 'shift595' is called. When the digit has been sent, the 74xx595 latch is pulsed to place the 74xx595 shift register content on to the output pins.

The subroutine converts the digit to segment codes with a **lookup** command, and then shifts the segment codes into the 74xx595 one bit at a time. The data output port is set high or low depending on the value of the bit to be sent. The shift register is then clocked to shift all bits into the next position and store the bit at the data output port in the first position of the shift register.

```
'Display hexadecimal numbers 00 to FF on 7 segment display using a serial interface
#picaxe 08m


symbol dataout = 4             'Define port for data stream
symbol clk = 0                 'Define port for clock
symbol latchout = 2            'Define port for latch

symbol bitcounter = b1         'Define variable names
symbol outbit = b2
symbol outbyte = b3
symbol counter = b4
symbol outdigit = b5


    do
        for counter = 0 to $FF     'Count to hexadecimal FF (decimal 255)
            outdigit = counter & $0F   'Isolate least significant digit
            gosub shift595         'Convert to segment codes and shift into 74xx595
            outdigit = counter / 16    'Isolate most significant digit
            gosub shift595         'Convert to segment codes and shift into 74xx595
            pulsout latchout, 1    'All digits have been sent, latch the data
            pause 100
        next counter
    loop
```

'Subroutine to convert a digit to segment codes and shift them into a 74xx595, 'the digit to be sent is in 'outdigit'

```
shift595:
    lookup outdigit, ($BE, $82, $DC, $D6, $E2, $76, $7E, $92, $FE, $F2, $FA, $6E,
        $3C, $DE, $7C, $78), outbyte
```

```
for bitcounter = 0 to 7          'Count to 8
   outbit = outbyte & 1          'Isolate the bit to be sent
   if outbit = 1 then            'test the bit, if its a 1 then
      high dataout               'set the data out port high
   else                          'otherwise
      low dataout                'set the data out port low.
   endif
   pulsout clk, 1                'Clock the shift register
   outbyte = outbyte / 2         'Shift right to position the next bit for output
next bitcounter
return
```

This code is very similar to the code for the single-digit serial interface. It uses an extra variable for the digit being displayed and two digits are sent.

Liquid Crystal Displays

LCDs have a number of advantages over LED displays in that they consume less power and are more cost-effective on a per-character basis. Common configurations for LCDs are one or two lines of 16, 20, or 40 characters, although other configurations are possible. Many LCD modules are based on the Hitachi 44780 controller and are relatively easy to interface to PICAXE chips.

Hitachi 44780-based LCDs have an 8-bit parallel interface, which can be software-configured for four bits. They typically have a connector consisting of 14 or 16 pins that are arranged in single row or two rows. The extra two pins on the 16-pin units are the connections for a backlight and these pins may be present even if the unit is not equipped with a backlight.

The pinout for 14- or 16-pin straight-line devices is shown in Fig. 7.31.

In operation, data is placed on the LCD data lines along with the instruction/register and read/write signals. The latch is then pulsed to strobe the data into the LCD. The LCD controller does all the work required to display the character or execute the instruction. LCDs have a busy flag that can be read by the processor to determine when the display is ready to receive data or instructions; however, in most applications, it is usually easier to ignore the busy flag and simply wait the maximum amount of time for each operation to complete before writing to the display again.

LCDs will operate in eight- or four-bit mode. In eight-bit mode, eight data bits are placed on the eight data lines and then latched into the LCD. In four-bit mode, a two-step process is used. First, the high-order four bits are placed on the high-order data lines and latched; then the low-order four bits are placed on the high-order data lines and latched. The four low-order data lines are not used in four-bit mode.

A minimum of six data lines are required to connect an LCD using four-bit mode and a minimum of 10 data lines using eight-bit mode. Using four-bit mode has some definite advantages when using the smaller PICAXE chips, which do not have 10 output

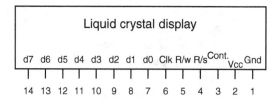

Pin 1 – Ground
Pin 2 – Vcc + 5 volts
Pin 3 – Contrast (0 to 5 volts)
Pin 4 – R/S Register / Instruction select
Pin 5 – R/W Read / Write select
Pin 6 – Latch or Enable
Pins 7 – 14 Data
Backlight pins may be adjacent to pin 1 or
pin 14, or located elsewhere.

Figure 7.31 Pin connections
for Hitachi 44780-based LCD.

lines available, although there is the disadvantage of slower operation because of the two-step process involved. The PICAXE-08M does not have sufficient output lines to be connected directly to an LCD, although it can be connected to a display by using an I/O expander or to an AXE033 serial LCD module.

To reduce the number of I/O lines required to address LCDs, Revolution Education make a serial LCD module available as part AXE033. This module is capable of operating in asynchronous serial mode requiring only one data line, or I2C mode requiring two data lines. There is also a stand-alone serial to LCD interface chip available as part FRM010.

The following circuits show three methods of interfacing a PICAXE to a Hitachi 44780-based LCD. The first circuit, suitable for use with all PICAXE chips, uses a shift register and requires three PICAXE output pins. The second circuit connects directly to an LCD operating in four-bit mode; it requires six PICAXE output pins. The third circuit uses the AXE033 module in both asynchronous serial and I2c modes.

Connecting an LCD Using a Shift Register

LCD modules must be initialized before they can be used. The initialization sequence varies depending on the data mode.

The initialization sequence for a two-line display used in eight-bit mode is:

1 Wait 15 ms or more after the display has been powered up.
2 Send $30 and wait 5 ms or more.
3 Send $30 and wait 160 μs or more.
4 Send $30 and wait 160 μs or more.

The display is now initialized and commands can be sent. A typical command sequence is:

1 Set the interface length ($38 for eight-bit, two lines, 5 × 7 characters).
2 Set scrolling ($10 for scrolling off).
3 Clear the display ($01).
4 Set cursor move direction ($06 for right, $04 for left).
5 Enable display/cursor ($0C for display on, cursor off, blink off).

Characters and further commands can now be sent to the display.

The initialization sequence for a two-line display being used in four-bit mode is:

1 Wait 15 ms or more after the display has been powered up.
2 Send $3 to the four high-order data lines and wait 5 ms or more.
3 Send $3 to the four high-order data lines and wait 160 μs or more.
4 Send $3 to the four high-order data lines and wait 160 μs or more.
5 Send $2 to the four high-order data lines and wait 5 ms or more.

The display is now initialized and commands can be sent. A typical command sequence is:

1 Set the interface length ($28 for four bit, two lines, 5 × 7 characters).
2 Set scrolling ($10 for scrolling off).
3 Clear the display ($01).
4 Set cursor move direction ($06 for right, $04 for left).
5 Enable display/cursor ($0C for display on, cursor off, blink off).

Characters and further commands can now be sent to the display as two four-bit writes with the high-order bits being sent first.

To write a character at the current cursor position, set R/S to 1 and R/W to 0, send the character and pulse the clock line high and then low. To send a command to the display, set R/S to 0 and R/W to 0, send the command, and pulse the clock line high and then low.

Some commonly used commands are:

Clear display	$01
Position cursor	$80 for line 1 + character position (beginning at zero)
	$C0 for line 2 + character position (beginning at zero)
Cursor off	$0C
Cursor on	$0E
Cursor on blinking	$0F
Cursor direction left	$04
Cursor direction right	$06

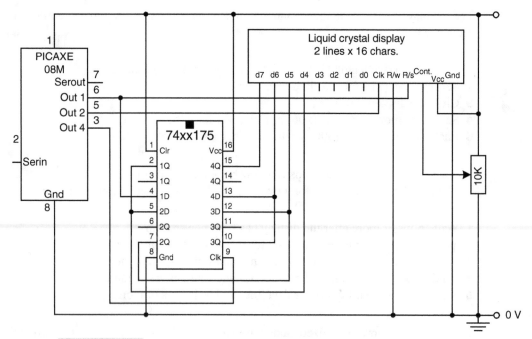

Figure 7.32 LCD in four-bit mode using a shift register for storage.

Figure 7.32 shows a practical circuit for the connection of an LCD module using a shift register. Figure 7.33 shows a practical circuit for the connection of an LCD module using direct connections. Figure 7.34 shows the connection for the AXE033 serial LCD in both serial and I2C modes.

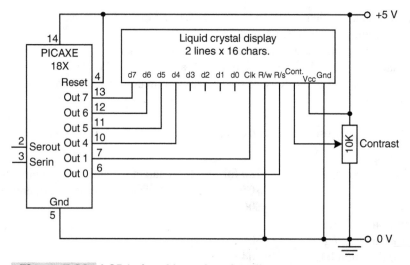

Figure 7.33 LCD in four-bit mode using direct connections

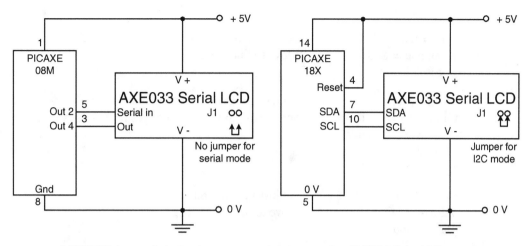

a. AXE033 in serial mode   b. AXE033 in I2C mode

Figure 7.34   Using the AXE033 in serial and I2C modes. (a) AXE033 in serial mode; (b) AXE033 in 12C mode.

The circuit of Fig. 7.32 uses the LCD in four-bit mode with a shift register to expand the number of PICAXE outputs. The code commences by defining symbols for constants and variables and then initializes the display to four-bit mode, cursor off, and cursor moving to the right.

```
'LCD four-bit interface using shift register
#picaxe 08m

symbol dataout = 1
symbol latchout = 2
symbol clk = 4

symbol lcddata = 1
symbol lcdinst = 0

symbol outbyte = b0          'Must be assigned to b0 for bit testing
symbol rs = b2
symbol bitcount= b3
symbol counter = b4
symbol niblcount = b5

                             'Initialize LCD
    rs = lcdinst             'Set instruction mode
    for counter = 0 to 6     'Send initialization commands
        lookup counter, ($33, $32, $28, $10, $01, $06, $0C), outbyte 'Initialization
            sequence
        gosub lcdout
    next counter
```

```
do
    rs = lcdinst                      'Set instruction mode
    outbyte = $80                     'Move to line 1 char 0
    gosub lcdout

    rs = lcddata                      'Set data mode and write letters A to O
    for counter = 65 to 79            'Set counter to the ASCII codes for letters A to O
        outbyte = counter
        gosub lcdout
    next counter

    rs = lcdinst                      'Set instruction mode
    outbyte = $C0                     'Move to line 2 char 0
    gosub lcdout

    rs = lcddata                      'Set data mode and write numbers 0 to 9
    for counter = 48 to 57            'Set counter to ASCII codes for numbers 0 to 9
        outbyte = counter
        gosub lcdout
    next counter

    loop                              'Repeat indefinitely
                                      'Subroutine to write to LCD in 4 bit mode with
                                      'shift register. Data to be written is in "outbyte"

lcdout:
    for niblcount = 1 to 2            'Count2 nibbles
        for bitcount = 0 to 3         'Count4 bits
            if bit7 = 1 then          'test the bit, if its a 1 then
                high dataout          'set the data port high
            else                      'otherwise
                low dataout           'set the data port low
            endif
            pulsout clk, 1            'Clock the shift register

            outbyte = outbyte * 2     'Shift right to position the next bit for output
        next bitcount

                                      'The byte has been sent to the LCD now set the mode
        if rs = lcddata then          'Test the mode and if its "data" then
            high dataout              'set the r/s line high (dataout has a dual function)
        else                          'otherwise it must be an instruction so
            low dataout               'set the r/s line low
        endif
        pulsout latchout, 1           'Pulse the LCD latch
        pause 4
    next niblcount
return
```

The circuit of Fig. 7.33 uses the LCD in four-bit mode with a direct connection to PICAXE. The code commences by defining symbols for constants and variables and then initializes the display to four-bit mode, cursor off, and cursor moving to the right.

```
'LCD four-bit interface using direct connection
#picaxe 18x

symbol latchout = 1
symbol lcddata = 1
symbol lcdinst = 0

symbol outbyte = b0
symbol rs = b1
symbol counter = b2
symbol nblcount = b3

    rs = lcdinst                'Set instruction mode
    for counter = 0 to 5        'Send initialization commands
       lookup counter, ($33, $32, $28, $0C, $01, $06), outbyte 'Initialization sequence
       gosub lcdout
    next counter

    do
       rs = lcdinst             'Set instruction mode
       outbyte = $80            'Move to line 1 char 0
       gosub lcdout

    rs = lcddata                'Set data mode and write "PICAXE-18X ABCDE"
    for counter = 0 to 15
       lookup counter ("PICAXE-18X ABCDE"), outbyte
       gosub lcdout
    next counter

    rs = lcdinst                'Set instruction mode
    outbyte = $C0               'Move to line 2 char 0
    gosub lcdout

    rs = lcddata                'Set data mode and write "1234567890"
    for counter = 0 to 9
       lookup counter ("1234567890"), outbyte
       gosub lcdout
    next counter

    loop                        'Repeat, indefinitely

                                'Subroutine to write to LCD in fourbit mode with
                                'direct connection. Data to be written is in "outbyte"
```

```
lcdout:
    for nblcount = 1 to 2
        pins = outbyte & $F0 | 2 | rs      'Outbyte + latch high + data/inst
        low latchout
        pause 4                            'Delay to enable LCD to execute command
        outbyte = outbyte * 16             'Shift left to position next nibble for output
    next nblcount
    return
```

Using the AXE033 serial and I2C modes is shown in Fig. 7.34.
Sample code for using the AXE033 in serial mode is:

```
'LCD using AXE033 in serial mode
#picaxe 08m

        pause 500                          'Delay to allow AXE033 module to initialize
        b0 = 123

        serout 2, N2400, ($FE, $80)        'Move to line 1 char 0
        serout 2, N2400, ("ABCDEFGHIJK")   'Write some letters

        serout 2, N2400, ($FE, $C1)        'Move to line 2 char 1
        serout 2, N2400, ("1234567890 ", #b0)  'Write some numbers and a variable
```

Code analysis

Instructions and data are sent to the AXE033 using the **serout** command and the code
is very much simpler than the code for a parallel connected LCD module. Commands
must be preceded by the character 254 ($FE), text must be enclosed in double quotes,
and variables will be treated as an ASCII code unless they are preceded by a # character,
in which case they will be converted to ASCII.
    Sample code for using the AXE033 in I2C mode is:

```
'LCD using AXE033 in I2C mode
#picaxe 18x
#no_data

        pause 500                          'Delay to allow AXE033 module to initialize
        b0 = 65

        i2cslave $C6,i2cslow,i2cbyte       'Initialize I2C, AXE033 device = $C6
```

```
writei2c 0, ($FE, $80, $FF)          'Move to line 1 char 0
pause 10                             'Delay to allow command to complete

writei2c 0, ("ABCDEF", $FF)          'Write some letters
pause 10                             'Delay to allow command to complete

writei2c 0, ($FE, $C1, $FF)          'Move to line 2 char 1
pause 10                             'Delay to allow command to complete

writei2c 0, ("123456", b0, $FF)      'Write some numbers and a variable
pause 10                             'Delay to allow command to complete
```

Code analysis

This code is very much simpler than the code for a parallel connected LCD module, because the AXE033 module takes care of all the initialization and data transfer to the display. The i2cslave command must be issued to initialize I2C for the AXE033; instructions and data are then sent using the **i2cwrite** command. Commands must be preceded by the character 254 ($FE), text must be enclosed in double quotes, and variables will be treated as an ASCII code. Note that a $FF character must be sent as the last character of a write command.

Keypad Input

In this experiment, a 4 × 4 matrix keypad is connected to a PICAXE -14M and key presses are displayed on a single-digit seven-segment LED display that is connected via a shift register. Output pin 2 of the PICAXE performs a dual function: it acts as the column-4 strobe and the clock for the shift register. This does not produce a conflict, even though the shift register will receive unexpected clock pulses, because the output latch of the shift register is not strobed and, thus, the output of the shift register will remain constant irrespective of the data in the shift register.

The circuit for the Matrix keyboard is shown in Fig. 7.35.

Matrix keypads are organized in rows and columns with a switch at the intersection of each row and column. To read a matrix keypad, it is necessary to set a column to the high state and then read all the rows to determine if a key in that column is pressed. The identity of a key can be determined by knowing which column is active and which row returns a result. This is demonstrated by the following segment of pseudocode.

```
Set column 1 to high
Read all rows
If row 1 = 1 then "1" key was pressed
If row 2 = 1 then "4" key was pressed
If row 3 = 1 then "7" key was pressed
If row 4 = 1 then "*" key was pressed
```

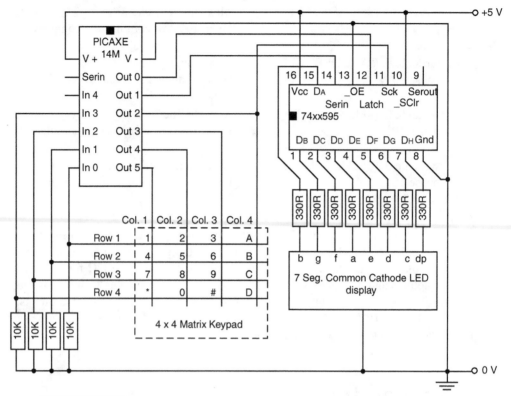

Figure 7.35 Circuit for a Matrix keypad.

Set column 2 to high
Read all rows
If row 1 = 1 then "2" key was pressed
If row 2 = 1 then "5" key was pressed
If row 3 = 1 then "8" key was pressed
If row 4 = 1 then "0" key was pressed

Columns 3 and 4 are a treated in a similar way.
The code to read the keypad and display the result is:

```
'Matrix keypad 4x4
#picaxe 14m

symbol dataout = 1          'Define port for data stream
symbol clk = 2              'Define port for clock
symbol latchout = 0         'Define port for latch

symbol bitcounter = b0      'Define variable names symbol outbit = b1
symbol outbyte = b2
symbol rowvalue = b3
symbol dispchar = b4
```

```
dispchar = 0                    'Initialize display character to blank

do

pins = $20         'Set column 1 high
lookup pins, (dispchar, $82, $E2, 0, $92, 0, 0, 0, $7C), dispchar '1, 4, 7, * (F)

pins = $10         'Set column 2 high
lookup pins, (dispchar, $DC, $76, 0, $FE, 0, 0, 0, $BE), dispchar '2, 5, 8, 0

pins = $08         'Set column 3 high
lookup pins, (dispchar, $D6, $7E, 0, $F2, 0, 0, 0, $78), dispchar '3, 6, 9, # (E)

pins = $04         'Set column 4 high
lookup pins, (dispchar, $FA, $6E, 0, $3C, 0, 0, 0, $DE), dispchar 'A, B, C, D

outbyte = dispchar
gosub shift595            'Shift into 74xx595

loop
```

'Subroutine to shift character into 74xx595 and display on seven-segment display
'the digit to be sent is in "outbyte"

```
shift595:
    for bitcounter = 0 to 7         'Count to 8
        outbit = outbyte & 1        'Isolate the bit to be sent
        if outbit = 1 then          'test the bit, if it's a 1 then
            high dataout            'set the data out port high
        else                        'otherwise
            low dataout             'set the data out port low.
        endif
        outbyte = outbyte / 2       'Shift right to position the next bit for output
    next bitcounter
    pulsout latchout, 1             'Latch the data
    return
```

Telephone Intercom

Here's a simple circuit that allows two telephones to be connected together. When either phone is picked up, the other will ring until it is answered. You can then talk for as long as you want for free. The circuit will work with almost any touch-tone telephone, including cordless phones. The circuit was originally designed and built for my children

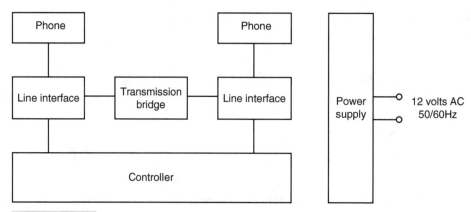

Figure 7.36 Block diagram of a telephone intercom.

to play with. They've grown up now and have completely forgotten about it in favor of MP3 players, mobile phones, laptop computers, and cars, but there must be other children in this world who would enjoy this and so the project remains. A project like this also has many other uses and it is a useful explanation, even if somewhat simple, of how a fixed line phone works and the logic behind making a connection between two phones.

To connect two phones, we need a circuit that can simulate the signals sent and received by a telephone exchange, and this is not difficult to do. Figure 7.36 shows a block diagram of what is needed. A line interface must be set up for each phone and a transmission bridge is needed to connect the speech signals. A controller determines when to ring the phones and the power supply provides all the voltages necessary to make the circuits work. The final circuit in Fig. 7.37 may look a little complicated at first, but it is really quite simple when it is broken down into its component parts.

The line interface is reproduced by itself for clarity in Fig. 7.38 with the components numbered. When the relay is in the normally closed position, power is supplied to the line via R1 and R2. With a supply of 30 V the line current will be around 25 mA, depending on line resistance and the type of phone. When the relay is in the operated position, ring current flows in the line via R3 and R2 and the phone rings. The off-hook condition is detected by measuring the line current that appears as a voltage across R2. Zero volts across R2 means that no current is flowing and, therefore, the phone is on hook. When the phone is off hook, there will be up to 12 V across R2, depending on line resistance and the type of phone.

The circuit formed by R4, R5, R6, R7, C1, and Q1 performs the off-hook detection. R4, R5, and R6 form a voltage divider network across R2; their combined resistance is high enough to have no effect on the line current. C1 in conjunction with R4 and R5, filters out unwanted AC components that may be present in the line current. The voltage at the junction of R5 and R6 is used to operate Q1. When line current flows, Q1 conducts and should become saturated; when no line current flows, Q1 is cut off. The voltage at the collector of Q1 is compatible with TTL levels and is fed directly to one of the input ports of the PICAXE.

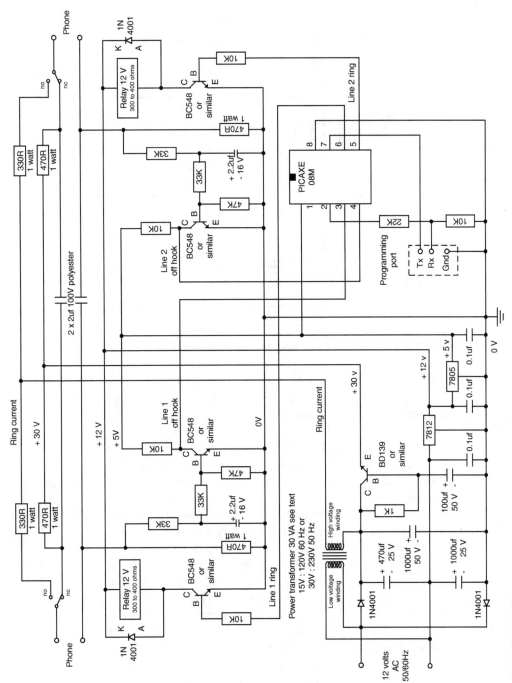

Figure 7.37 Circuit of a telephone intercom.

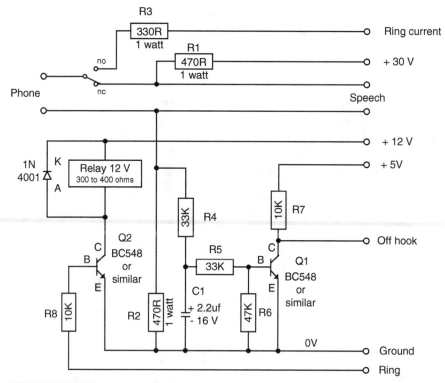

Figure 7.38 Telephone intercom line interface circuit.

A TTL-level signal from the PICAXE will operate the ring relay via Q2 and R8. The 1N4001 diode across the relay protects Q2 from back EMF when the relay releases. The speech signal appears directly across the speech terminals of the line interface and is connected to the transmission bridge. Some of the speech signal will also pass through R1 and R2 and be lost. In a simple circuit, such as this where lines are short, this is not a problem and there will be sufficient speech signal available for the other phone. The transmission bridge can be seen in the complete circuit. It consists of two 2-μF capacitors that allow the speech signal to pass from one line interface to the other while blocking DC.

The controller can be seen in the complete circuit. It consists of a PICAXE-08M microcontroller that is programmed to read the status of the off-hook signals from each phone and then use logic to provide ring signals to the phones.

The power supply is reproduced with the components numbered in Fig. 7.39. The power supply operates from a 12-V AC input that can be conveniently supplied from a plugpack (wall transformer). The incoming AC is rectified by a voltage doubler circuit consisting of D1, D2, C1, C2, and C3 to produce about 32 V across C3. One-half the power supply output is taken from C2 and fed into a 12-V regulator for the relay supply and then a 5-V regulator for the microcontroller supply. The 0.1 μF capacitors around the regulators prevent parasitic oscillations and should be monolithic bypass capacitors.

Capacitor C3 acts as a reservoir for the DC voltage and provides a low-impedance path to ground for the ring current. Any ripple present across C3 will produce an

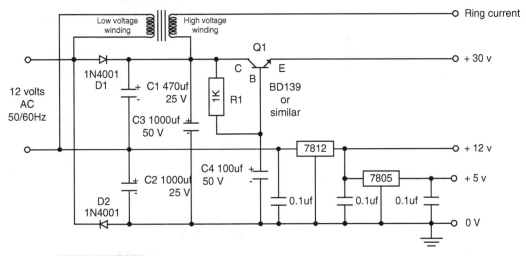

Figure 7.39 Telephone intercom power supply.

annoying hum in the telephones. To overcome this, R1, C4, and Q1 are used to form an active filter circuit, reducing the hum to an inaudible level. The filter will reduce the DC output to around 30 V and this is sufficient to operate most phones; the actual voltage is not critical and any voltage between 24 and 48 should work.

The ring current is supplied by a mains transformer used in reverse to provide about 90 V AC to ring the phones. The transformer specified has a voltage ratio of 8:1 and this will produce about 96 V from a 12-V input. The actual voltage is not critical and any voltage between 70 and 100 V should work, although 100 V should be considered an absolute maximum. There is sufficient voltage here to give an unpleasant shock—so keep your fingers clear! During ringing, DC flows through the transformer primary (high-voltage winding) and the resistance of this winding should not exceed 140 Ω in order to ensure that the off-hook detection circuitry operates correctly. A 30 VA transformer should meet this requirement. The 330-Ω resistor (R3) in the line interface should not be reduced in value as it limits the current that can flow in the event of a short circuit on the line.

Resistors R1, R2, and R3 can get hot when operating and they should be mounted at least 6 mm (1/4 in) above the surface of the circuit board to enable air to circulate around them. Alternately, 5-watt resistors can be used to keep their temperature lower.

TESTING

Ensure that no phones are plugged in and that the microcontroller is not in circuit. Apply 12 V AC and measure the power supply voltages. The DC should be around 30 to 32 V; the exact value is not critical and it may vary a little depending on mains voltage. The relay supply should be 12 V and the microcontroller supply should be

5 V. Use an AC range on a multimeter to measure the ring voltage between the two high-voltage terminals on the transformer; it should be between 70 and 100 V AC. Be careful doing this because there is sufficient voltage to give an unpleasant shock. When the power supply voltages have been verified, disconnect the 12-V AC supply and wait for the capacitors to discharge, connect the microcontroller, and plug a phone into each socket. Reconnect the 12-V AC supply and apply power to the circuit. Lift one phone and verify that the other phone rings. When it does, pick it up and verify that the ring stops. If the ring continues, it will be clearly heard as a loud buzzing noise. Be careful not to put the phone to your ear until you verify that the ringing has stopped. A burst of a ring in your ear can be painful! Repeat the process using the other phone and verify that you can talk from one phone to the other.

IF IT DOESN'T WORK

If the power supply voltages are not as they should be, carefully check the power supply for wiring faults, wrong components, short or open circuits, and component polarity. Try checking the voltages across C1, C2, and C3. If a phone doesn't ring, first check both line circuits. Use a multimeter to measure the voltage between the phone terminals with the phone disconnected. It should be nearly the same as the DC output of the power supply. If that's okay, then measure the voltage across R2 with the phone on and off the hook. It should be zero when the phone is on hook and between 6 and 12 V when the phone is off the hook. Measure the voltage at the collector of Q1 with the phone on and off the hook. It should be 5 V with the phone on hook and less than 0.5 V with the phone off hook. To test ringing, disconnect power and wait for the capacitors to discharge. Remove the PICAXE and reconnect power with both phones connected. Now, using a jumper lead, temporarily connect the ring terminal for each line to +5 V. Verify that the relay operates and the phone rings. If you are having trouble with a phone, try substituting another phone that is known to be in working order. Note that older phones equipped with mechanical bells may not work properly with this circuit.

CONNECTING PHONE LINES

The circuit operates with two wire-balanced lines and is not suitable for use with single-wire, earth return. Just about any type of wire can be used so long as it has two conductors. Telephone cable is an obvious choice, but speaker wire will work well as will fencing wire, providing the insulation between the wires does not allow leakage current to flow in the line. Wet fence posts do not make good insulators and it's probably a good idea to keep phone wiring away from electric fence wiring. The circuit should work with lines several hundred meters (yards) in length depending on the type of wire used. The line resistance will be the main factor determining the length of the line. Standard 0.5-mm diameter phone wire should be okay up to 200 mm (yards) or so. It may be necessary to use thicker wire for longer distances.

It would be a good idea not to use standard telephone connectors so that you will not accidentally connect this circuit to the public network. In practice, you must not connect this circuit to the public phone network, since it is a breach of laws or regulations to do so and offenders may face heavy fines. There is also no point in connecting it to the public phone network because it will not work. If you want to connect something to the public network, just use a phone (they're designed for that purpose). Note that laws or regulations may prevent you from extending the lines from this circuit outside the boundaries of your own property.

```
'Telephone intercom
#picaxe 08m

symbol line1ring = 1
symbol line2ring = 2
symbol line2offhook = pin3
symbol line1offhook = pin4
symbol no = 1
symbol yes = 0

symbol ringinhibit = b0

    do
    if line1offhook = no and line2offhook = no then
        low line1ring
        low line2ring
        ringinhibit = no

        elseif line1offhook = no and line2offhook = yes and ringinhibit = no then
        high line1ring

        elseif line1offhook = yes and line2offhook = no and ringinhibit = no then
        high line2ring

        elseif line1offhook = yes and line2offhook = yes then
        low line1ring
        low line2ring
        ringinhibit = yes

    endif
    loop
```

The code is reasonably simple. The most complicated part is the logic that is needed to disconnect the ring when the phones are off hook. The **ringinhibit** variable is used to prevent a phone ringing at the end of a call until both phones are on hook.

Voltmeter

This experiment shows a method of using an analog input to measure voltage and display it on an LCD. Figure 7.40 shows the arrangement; Fig. 7.41 shows the circuit.

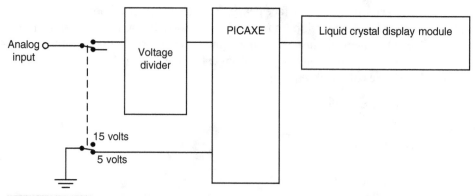

Figure 7.40 Arrangement for a voltmeter.

CIRCUIT DESCRIPTION

The circuit contains four main components: a voltage divider, a range switch, a PICAXE-08M, and a serial LCD module.

For the 5-V range, the input is connected directly to the PICAXE analog input pin. For the 15-V range, the input voltage is connected to a voltage divider that divides by 3. This is a very simple input circuit that does not have any overvoltage or built-in reverse polarity protection. The user should take care not to apply reverse polarity or excess voltage to the input terminal in order to avoid damaging the PICAXE chip.

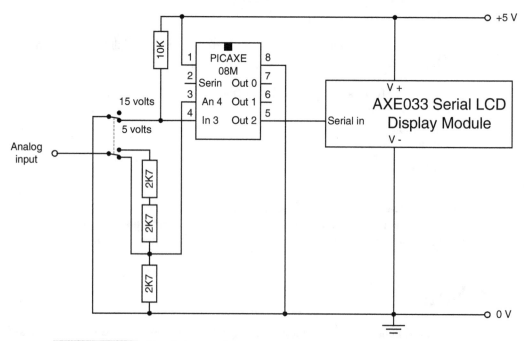

Figure 7.41 Circuit for a voltmeter.

The input impedance of the circuit will be approximately 2,700 Ω for the 5-V range and 8,100 Ω for the 15-V range. The range switch is a DPDT toggle switch. One pole is used to switch the input to the voltage divider and the other pole is used to inform the PICAXE which range is selected.

The display is a serial LCD module (part AXE033).

The code to read a voltage and display voltage on an AXE033 LCD module is:

```
'Analog voltmeter - dual range
#picaxe 08m

symbol rangemultiplier = 3          'Multiplier for 15-V range
symbol rangeswitch = pin3           'Range switch port
symbol range5volt = 0               '5-V range
symbol analogport = 4               'Analog input port
symbol displayport = 2              'Serial LCD port

symbol outbyte = b0                 'Output data
symbol analogdata = w3              'Analog reading - word
symbol analoglo = b6                'Analog reading - low byte
symbol analoghi = b7                'Analog reading - high byte

        pause 500                                    'Wait for AXE033 to initialize
        serout displayport, N2400, (254, 1, 254, $04)  'Clear display and set cursor
                                                            moves left


        do

        'Read 8 bit analog value and convert to a word for 16-bit arithmetic
        readadc analogport, analoglo    'Read eight-bit analog value into low byte of word
        analoghi = 0                    'Clear high byte of word

        'Convert to voltage and apply the multiplier if a voltage divider is being used
        analogdata = analogdata * 196 / 100   'Convert analog to volts with two decimal
                                                    places

        if rangeswitch <> range5volt then           'Test range switch setting
            analogdata = analogdata * rangemultiplier  '15-V range, apply range multiplier
        endif

        'Set the LCD cursor to receive the least significant digit
        serout displayport, N2400, (254, 133)       'Move to line 1, char 5

        'Convert the binary voltage to ASCII
        outbyte = analogdata // 10 + $30   'Isolate hundredths digit and convert to
                                                ASCII

        analogdata = analogdata / 10       'Remove hundredths
```

serout displayport, N2400, (outbyte)	'Display the hundredths digit
outbyte = analogdata // 10 + $30	'Isolate tenths digit and convert to ASCII
analogdata = analogdata / 10	'Remove tenths
serout displayport, N2400, (outbyte)	'Display the tenths digit
serout displayport, N2400, (".")	'Display the decimal point
outbyte = analogdata // 10 + $30	'Isolate units digit and convert to ASCII
analogdata = analogdata / 10	'Remove the units digit
serout displayport, N2400, (outbyte)	'Display the units digit
outbyte = analogdata + $30	'Remaining value is tens digit; convert it to ASCII
serout displayport, N2400, (outbyte)	'Display the tens digit
loop	'Repeat indefinitely

Code description

The program enters a permanent loop that reads the analog port, reads the range switch, converts the analog reading to a voltage, and displays the voltage on the LCD.

The analog value is read as an 8-bit binary number and is converted to a 16-bit number to prevent arithmetic overflow during calculations. The range 0 to 255 has 256 distinct values, and each step of the analog reading thus represents 19.6078 mV for the 5-V range and 3×19.6078 mV for the 15-V range. The overall accuracy of the voltage readings is determined by the amount of error in the analog-to-voltage conversion and is around 4% low.

1-Wire Serial Number

Each 1-wire device has a unique 64-bit serial number written into it at the time of manufacture which is composed of an 8-bit family code, a 48-bit serial number, and an 8-bit cyclic redundancy check (CRC).

This experiment reads the serial number from a 1-wire device and displays the result on an LCD. The display is in hexadecimal with the CRC and family code on line 1 and the serial number on line 2 in the same format as the serial number engraved on an iButton. Any 1-wire device may be used with appropriate connections. In this circuit, only one device can be connected to the 1-wire bus at a time. The circuit is shown in Fig. 7.42

The code to read a 1-wire serial number and display in on a serial LCD is:

```
'One-wire serial number
#picaxe 08m
```

symbol owport = 4	'Define the port for the 1-wire device
symbol lcdport = 2	'Define the port for the serial LCD

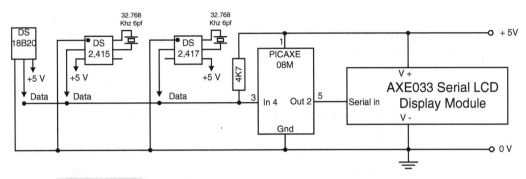

Figure 7.42 Circuit to read a 1-wire device serial number.

```
symbol hexdigit = b0
symbol asciichar = b1
symbol counter = b2
symbol niblcount = b3
symbol currentbyte = b5          'Current byte when processing serial number
symbol owfamily = b6             'readowsn command puts family code in b6
symbol owser6 = b7               'Serial number, least significant byte
symbol owser5 = b8               'Serial number
symbol owser4 = b9               'Serial number
symbol owser3 = b10              'Serial number
symbol owser2 = b11              'Serial number
symbol owser1 = b12              'Serial number, most significant byte
symbol owchecksum = b13          'readowsn command puts checksum in b13

    pause 500                    'Wait for AXE033 to initialize
    serout lcdport, N2400, (254, 1)   'Clear display
    pause 30


    do
    readowsn owport              'Read the serial number
    if owfamily <> 0 then        'If a 1-wire device is present

    serout lcdport, N2400, (254, $8A)    'Position cursor and display the
                                            family code

      currentbyte = owfamily
      for niblcount = 0 to 1
        hexdigit = currentbyte / 16       'Isolate and align digit
        lookup hexdigit,                  'Convert to ASCII
          ("0123456789ABCDEF"), asciichar
        serout lcdport, N2400, (asciichar)
        currentbyte = currentbyte * 16    'Position lower nibble for output
      next niblcount
```

```
    serout lcdport, N2400, (254, $80)              'Position cursor and display the
                                                        CRC

        currentbyte = owchecksum
        for niblcount = 0 to 1
            hexdigit = currentbyte / 16            'Isolate and align digit
            lookup hexdigit,                       'Convert to ASCII
                ("0123456789ABCDEF"), asciichar
            serout lcdport, N2400, (asciichar)
            currentbyte = currentbyte * 16         'Position lower nibble for output
        next niblcount

    serout lcdport, N2400, (254, $C0)              'Position cursor and display the
                                                        serial number

    for counter = 1 to 6
        currentbyte = owser1
        for niblcount = 0 to 1
            hexdigit = currentbyte / 16            'Isolate and align digit
            lookup hexdigit,                       'Convert to ASCII
                ("0123456789ABCDEF"), asciichar
            serout lcdport, N2400, (asciichar)
            currentbyte = currentbyte * 16         'Position lower nibble for output
        next niblcount

        owser1 = owser2                            'Position remaining digits for output
        owser2 = owser3
        owser3 = owser4
        owser4 = owser5
        owser5 = owser6
    next counter

    owfamily = 0                       'Clear the family code
    else '(owfamily <> 0 then)
    serout lcdport, N2400, (254, 1)    'Clear display if no 1-wire device is present
    pause 30
    endif '(owfamily <> 0 then)
    loop
```

1-Wire Temperature

This experiment reads the temperature from a DS18B20 digital thermometer and displays it in degrees Celsius and Fahrenheit on an AXE033 LCD. There are two commands to read the temperature from a DS18B20, **readtemp** and **readtemp12.** Code is given for both commands and the Fahrenheit conversion routines are given for the **readtemp** command version.

A circuit for the 1-wire temperature experiment is shown in Fig. 7-43.

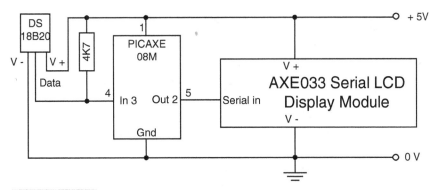

Figure 7.43 Circuit for a digital thermometer.

The **readtemp** command reads the temperature in degrees Celsius and places it into a byte. The temperature can be in the range +125 degrees to −55 degrees and the numeric portion is placed into the least significant seven bits of the byte (bits 6 – 0) as a positive number, i.e., 0 to 125 for positive temperatures and 0 to 55 for negative temperatures. The most significant bit of the byte is the sign indicator and is 0 for positive temperatures and 1 for negative temperatures.

The relationship between degrees Celsius and Fahrenheit is given by the expression:

$$(F - 32) / C = 9 / 5$$

where

F is the temperature in degrees Fahrenheit

C is the temperature in degrees Celsius

Transposing this expression gives the following two expressions for calculating Celsius from Fahrenheit and Fahrenheit from Celsius.

$$C = 5(F - 32) / 9$$
$$F = 9C / 5 + 32$$

A few equivalent temperatures are:

Celsius	Fahrenheit
100	212
0	32
−17.778	0
−40	−40

The following code uses the **readtemp** command and displays the temperature in degrees Celsius and Fahrenheit on an AXE033 serial LCD using the circuit in Fig. 7.43.

```
'One-wire temperature Celsius and Fahrenheit
#picaxe 08m

symbol owport = 4          'Define the port for the 1-wire device
symbol lcdport = 2         'Define the port for the serial LCD

symbol tempreading = b0    'Temperature reading
symbol csign = b1          'Sign (+ or −) for Celsius temperature reading
symbol fsign = b2          'Sign (= or −) for Fahrenheit reading
symbol fahreading = w3     'Work space for producing Fahrenheit reading
symbol fahdecdigit = b4    'Work space to hold decimal digit for Fahrenheit reading

    pause 500                           'Wait for display to initialize
    serout lcdport, N2400, (254, 1)     'Clear display
    pause 30

    do
    readtemp owport, tempreading        'Read the temperature
    gosub displayC                      'Display Celsius on line 1
    gosub displayF                      'Display Fahrenheit on line 2
    loop

displayC:                     'Display the temperature in Celsius
if tempreading > 127 then     'Test for the sign indicator in bit 7
    csign = "−"                   'If bit 7 is set then sign is "−"
    tempreading = tempreading & $7F   'Remove sign bit
else                              'otherwise
    csign = "+"                   'sign is "+"
endif

    serout lcdport, N2400, (254, $80)   'Move to line 1
                                        'Display sign, temperature, + some extra
                                            spaces
    serout lcdport, N2400, (csign, " ", #tempreading, "C")
return

displayF:       'Display the temperature in Fahrenheit
                'A single decimal digits is required for integer Celsius temperatures
    if tempreading > 127 then             'Test for the sign indicator in bit 7
        csign = "−"                       'If bit 7 is set then Celsius sign is "−"
        tempreading = tempreading & $7F   'Remove sign bit
    else                                  'otherwise
        csign = "+"                       'Celsius sign is "+"
    endif
```

```
fahreading = tempreading * 10          'Calculate Fahrenheit from Centigrade
                                       'Move temperature to word and scale 1
                                          decimal place

fahreading = fahreading * 9 / 5
if csign = "−" then
    fahreading = fahreading xor        'Complement for negative Celsius
        $FFFF + 1                         temperatures
    endif
fahreading = fahreading + 320
if fahreading > 32767 then             'Test for the sign indicator in bit 15
    fsign = "−"                        'Fahrenheit temperature is −ve
    fahreading = fahreading xor        'Complement
        $FFFF + 1
    else
    fsign = "+"                        'Fahrenheit temperature is +ve
    endif

fahdecdigit = fahreading // 10         'Isolate tenths digit
fahreading = fahreading / 10           'Remove tenths digit

serout lcdport, N2400, (254, $C0)  'Move to line 2
                             'Display sign, temperature, +some extra spaces
serout lcdport, N2400, (fsign, " ", #fahreading, ".", #fahdecdigit, "F ")
return
```

The **readtemp12** command reads the temperature in degrees Celsius and places it into a word as a 12-bit binary number with four binary places. Negative temperatures are stored as two's complement binary numbers with the sign in bit 12.

The following code uses the circuit in Fig. 7.43 to read a 12-bit temperature reading and display it on a serial LCD in degrees Celsius with four decimal places.

```
'One-wire temperature 12-bit Celsius
#picaxe 08m

symbol owport = 4                'Define the port for the 1-wire device
symbol lcdport = 2               'Define the port for the serial LCD

symbol tempreading = w0          'Temperature reading
symbol sign = b4                 'Sign (+ or −) for temperature reading
symbol dec1 = b5                 '0.1's digit
symbol dec01 = b6                '0.01's digit
symbol dec001 = b7               '0.001's digit
symbol dec0001 = b8              '0.0001's digit
symbol readingfrac = b9          'Fractional portion of reading

    pause 500
    serout lcdport, N2400, (254, 1)
    pause 30
```

```
do

  readtemp12 owport, tempreading          'Read the temperature
  if tempreading > 1023 then              'Test for the sign indicator in bit 12
      sign = "–"                          'Sign is "–"
                                          'Propagate sign into high-order bits and
                                             complement
      tempreading = tempreading or $F000 xor $FFFF + 1
  else                                    'otherwise
      sign = "+"                          'sign is "+"
  endif

  readingfrac = tempreading and $0F       'Isolate the fractional part
  tempreading = tempreading / 16          'Shift right four bits to remove fraction
                                    'Convert fractional part to decimal and ASCII
  readingfrac = readingfrac * 10             'Generate decimal digit in high-order
                                                four bits
  dec1 = readingfrac / 16 + $30           'Isolate 0.1's digit and convert to ASCII
  readingfrac = readingfrac and $0F * 10  'Remove high order digit, generate next
                                             digit
  dec01 = readingfrac / 16 + $30          'Isolate 0.01's digit and convert to ASCII
  readingfrac = readingfrac and $0F * 10  'Remove high order digit, generate next
                                             digit
  dec001 = readingfrac / 16 + $30         'Isolate 0.001's digit and convert to
                                             ASCII
  readingfrac = readingfrac and $0F * 10  'Remove high order digit, generate next
                                             digit
  dec0001 = readingfrac / 16 + $30        'Isolate 0.00001's digit and convert to
                                             ASCII

  serout lcdport, N2400, (254, $80)       'Move to line 1 position 1
                                          'Display sign, temperature, decimal point
                                             and places + some extra spaces
  serout lcdport, N2400, (sign," ", #tempreading, ".", dec1,dec01,dec001,
      dec0001,"C")
  pause 10

loop
```

Radio Frequency Identification

Radio frequency identification (RFID) has many different applications and is often used for secure entry systems and object tracking. At first, it may seem to be difficult to set up an RFID system, but the availability of modern integrated circuits makes the task quite simple.

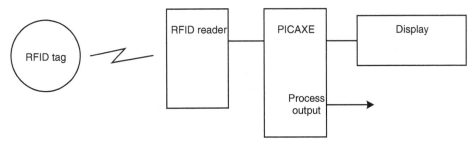

Figure 7.44 Block diagram of a RFID reader and display system.

RFID systems have two main components: readers and tags. Communicating between them can be done by inductive coupling or propagation coupling. In an inductive coupling system, the reader is an active device that generates a continuously alternating magnetic field at a specific frequency. The tag is a passive device that contains a pickup coil and an integrated circuit. When the tag is moved into the magnetic field of the reader, power is transferred to its pickup coil. The integrated circuit in the tag then sends information back to the reader using the same magnetic coupling loop from which it receives power. Some tags have facilities to receive and store information from the reader, while other tags are read-only.

Different frequency bands are in use for RFID devices and this experiment describes a system that uses inductive coupling and read-only tags operating in the low band at 125 KHz.

The experiment uses the PICAXE-18X to read the 10-digit number and the 2-digit checksum from an RFID tag and displays them on an LCD. A block diagram of the system is shown in Fig. 7.44; circuits are shown in Figs. 7.45 and 7.46.

The RFID reader used is the ID-12, which is a low-cost short-range proximity reader operating in the low band at 125 KHz. The ID-12 is self-contained and needs only a

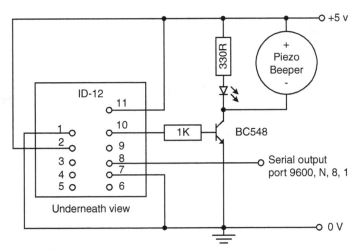

Figure 7.45 Circuit for a RFID reader.

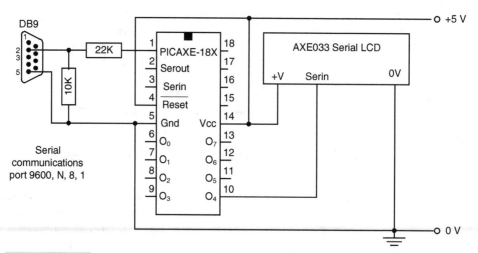

Figure 7.46 Circuit for RFID ID display.

+5-V power supply to operate. It communicates with the outside world by means of an internal RS232 serial port. Optionally, an LED and/or piezo beeper can be connected to the ID-12 via a driver transistor to provide the user with an indication of a successful read. The ID-12 is available in Australia from Adilam Electronics (www.adilam.com.au) and can be purchased as an individual chip or in an evaluation kit. The LCD is an AXE033 serial LCD module available from PICAXE distributors.

The final circuit is in two parts, the RFID reader, and the display. The circuits are shown in Figs. 7.45 and 7.46. The output from the ID12 may also be tested by connecting to a PC running a terminal program, such as Hyper Terminal. The communication parameters are 9,600 baud, no parity, eight data bits, and one stop bit.

The code to read and display the RFID ID on an LCD is so simple that it requires little explanation. Note, however, that the PICAXE runs at 8 MHz to allow serial communication at 9,600 baud and this doubles all baud rates, including those that address the serial display. All RFID ID messages begin with a start-of-message byte, which is hexadecimal 02. This value is used as a qualifier in the serin command. The SOM is followed by 10 bytes of hexadecimal characters, in ASCII format, and a 2-byte checksum. The checksum is followed by a carriage return, line feed, and Hex. 03, which are all ignored. This is documented by comments in the code.

The code to display the RFID ID using the circuit of Fig. 7.46 is:

```
'PICAXE-18X display RFID on LCD
#picaxe 18X

symbol lcdport = 4      'Define serial port for LCD
symbol rfidport = 2     'Define serial port for RFID
symbol line1 = $80      'Define line 1

    setfreq m8          'Run at 8 MHz - halves all delay times - doubles baud rates
```

```
pause 500                          'Wait for LCD to initialize
serout lcdport, N1200, (254, 1)    'Clear LCD
pause 30                           'Wait for command to complete

'Read RFID code and checksum
'Data format
'BOM $02
'RFID ID 10 ASCII characters ( 0 - F)
'Checksum 2 ASCII characters ( 0 - F)
'CR $0D
'LF $0A
'EOM $03

do

serin rfidport, N4800, ($02),b2,b3,b4,b5,b6,b7,b8,b9,b10,b11,b12,b13
serout lcdport, N1200, (254, line1)                'Move to line 1

'Display the RFID code and checksum

serout lcdport, N1200, (b2,b3,b4,b5,b6,b7,b8,b9,b10,b11," ",b12,b13)

loop
```

Simple ASCII Terminal

In this experiment, a PC keyboard and an LCD are used to make a simple ASCII terminal that operates at 2,400 or 9,600 bauds with either true or inverted transmission levels. The terminal has two distinct sections, the receiver and the transmitter, as shown in the block diagram in Fig. 7.47. The serial port for this circuit operates at TTL levels and a MAX232, or similar chip, can be used if true RS232 levels are required.

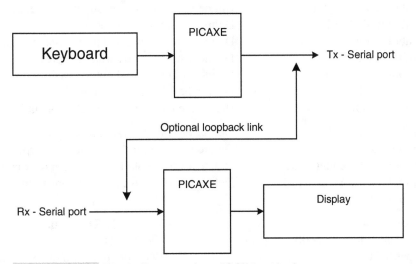

Figure 7.47 Block diagram of an ASCII terminal.

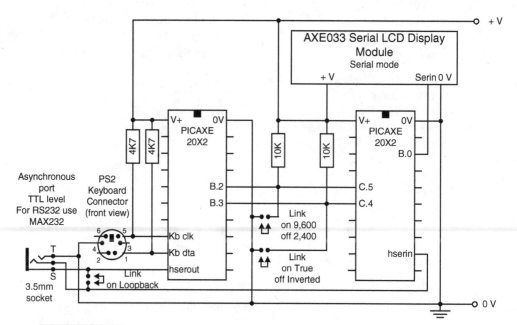

Figure 7.48 Circuit for an ASCII terminal.

The final circuit uses separate PICAXE-20X2 chips for the receiver and transmitter and is shown in Fig. 7.48. PICAXE-20X2 chips were chosen for their ability to invert transmission levels and to be able to receive in the background.

There are two code listings, one for the transmitter and one for the receiver.

```
'ASCII Terminal - Receive
#picaxe 20x2
```

symbol ratesel = pinC.5	'Define input pin for baud rate selection
symbol levelsel = pinC.4	'Define input pin for transmission level selection
symbol R2400 = 1	'Define a state for 2400 baud
symbol TxInv = 1	'Define state for inverted transmission level
symbol nodata = 0	'Define a state for no serial data present
symbol lcdport = B.0	'Define a pin for serial out to AXE033 LCD
symbol line1 = $80	'Define the LCD line 1 value
symbol lcdlen = 16	'Define the number of characters on an LCD line
symbol kbval = b0	'Define variable for value of keypress
symbol kbchar = b1	'Define variable for ASCII char of keypress
symbol lcdcharcount = b2	'Define a variable fo the LCD character counter
symbol lcdline = b3	'Define a variable for the LCD line pointer
symbol keymask = b4	
pause 500	'Wait for LCD to initialize
serout lcdport, N2400, (254, 1)	'Clear LCD
pause 30	'Wait for command to complete

```
lcdline = line1                              'Initialize LCD line pointer to line 1
serout lcdport, N2400, (254, lcdline)        'Move to line 1

'Setup hardware serial port speed and mode depending on jumpers
if ratesel = R2400 then                      'Test the rate select jumper
    if levelsel = TxInv then                 'Test the mode select jumper
        hsersetup B2400_8, %111              '2400 baud, inverted levels, background receive
    else
        hsersetup B2400_8, %001              '9600 baud, true levels, background receive
    endif '(if levelsel = TxInv)

else '(if ratesel = R2400)

                                             'Rate is 9600
    if levelsel = TxInv then                 'Test the mode select jumper
        hsersetup B9600_8, %111              '9600 baud, inverted levels, background receive
    else
        hsersetup B9600_8, %001              '9600 baud, true levels, background receive
    endif '(if levelsel = TxInv)

endif '(if ratesel = R2400)

setintflags $20, $20     'Set interrupt on serial background receive

do                       'Start an endless loop which does nothing,
                         'all processing is by interrupt. Program code can be placed
                         'here but it must not interfere with interrupts.
                         'E.g. kbin and other commands disable interrupts briefly
loop

interrupt:
    ptr = 0                                  'Set scratchpad pointer to 0 (1st char.)
    do while ptr < hserptr                   'While there are characters in the input buffer
        serout lcdport, N2400, (@ptrinc)     'Send char. to LCD and move to next char
        pause 5
        inc lcdcharcount                     'Increment character counter
        if lcdcharcount >= lcdlen then       'If the last character on the line is used then
            lcdline = lcdline xor $40        'move to next line
            serout lcdport, N2400, (254, lcdline)
            pause 5
            lcdcharcount = 0
        endif
    loop
    hserptr = 0                              'Reset the hardware serial pointer
    hserinflag = nodata                      'Reset the hardware serial flag
setintflags $20, $20                         'Reset interrupt on serial background receive
return                                       'Return from interrupt
```

```
'ASCII Terminal - Transmit
#picaxe 20x2

symbol ratesel = pinB.2      'Define input pin for baud rate selection
symbol levelsel = pinB.3     'Define input pin for transmission level selection
symbol R2400 = 1             'Define a state for for 2400 baud
symbol TxInv = 1             'Define state for inverted transmission level

symbol kbval = b0            'Define variable for value of keypress
symbol kbchar = b1           'Define variable for ASCII char of keypress
symbol keymask = b4

    'Setup hardware serial port speed and mode depending on jumpers
    if ratesel = R2400 then
        if levelsel = TxInv then
            hsersetup B2400_8, %111      '2400 baud, inverted levels, background receive
        else
            hsersetup B2400_8, %001      '9600 baud, true levels, background receive
            endif '(if levelsel = TxInv)
        else '(if ratesel = R2400)
        if levelsel = TxInv then
            hsersetup B9600_8, %111      '9600 baud, inverted levels, background receive
        else
            hsersetup B9600_8, %001      '9600 baud, true levels, background receive
            endif '(if levelsel = TxInv)
    endif '(if ratesel = R2400)

    do                           'Start and endless loop
    kbin [10, nokey], kbval      'Read a character from the keyboard
    read kbval, kbchar           'Convert to ASCII
        if kbchar > $07 then
            hserout 0, (kbchar)      'Send the character to the serial port
            pause 30                 'Pause to allow user to release the key (may be varied)
        endif
nokey:                           'Bypass keyboard read if no key pressed
    loop

#rem    Here's a useful piece of code that dumps the key code to an LCD in hex
            format.
        You will also need to connect an AXE033 LCD to B.0 and initialize it

symbol lcdport = B.0        'Define a pin for serial out to AXE033 LCD

        pause 500                       'Wait for LCD to initialize
        serout lcdport, N2400, (254, 1) 'Clear LCD
        pause 30                        'Wait for command to complete
```

```
        lcdline = line1                         'Initialize LCD line pointer to line 1
        serout lcdport, N2400, (254, lcdline)   'Move to line 1

keydump:        'Useful for debugging key codes. Send the key code to LCD in hex
                    format.
        kbchar = kbval / 16                     'Isolate high nibble and locate in low
                                                    order 4 bits
        lookup kbchar, ("0123456789ABCDEF"), kbchar
        serout lcdport, N2400, (kbchar)         'Send to LCD
        kbchar = kbval & $0F                    'Isolate low nibble
        lookup kbchar, ("0123456789ABCDEF"), kbchar
        serout lcdport, N2400, (kbchar, " ")    'Send to LCD followed by a space
        return
#endrem


'Keycode translation table
'A more complete list of key codes can be found in the PICAXE manual
'To decode "shift" characters the state of the shift and/or caps lock keys must be
    monitored
'and a separate translation table used for shifted characters
'The + sign is used for nonprinting chars


EEPROM $00,("+++++++++++++++'+")
EEPROM $10,("+++++Q1+++ZSAW2+")
EEPROM $20,("+CXDE43++ VFTR5+")
EEPROM $30,("+NBHGY6+++MJU78+")
EEPROM $40,("+,KIO09++./L;P-+")
EEPROM $50,("++'+[=+++++]++++")
EEPROM $60,("++++++++++1+4++++")
EEPROM $70,("0+2568++++3-*9++")
```

I2C Memory Expansion

Two 24LC256 I2C EEPROM chips are used to demonstrate the use of the I2C bus for memory expansion and the use of more than one chip on the bus. The 24LC256 chips have the low-order three bits of the chip address connected to pins, which can be used to provide eight unique addresses. In this experiment, the chips are connected at addresses 000 and 001. This is the low-order three bits of the address, which will be combined with the high-order four bits of the address that are configured at the time of manufacture. The overall addresses of the chips will be 1010 000x and 1010 001x.

The circuit is shown in Fig. 7.49.

'I2C memory expansion using two 24LC256

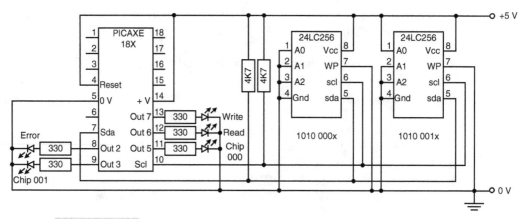

Figure 7.49 I2C memory expansion.

The code for I2C memory expansion is:

```
'I2C Memory expansion using 2 × 25LC256 - Test routine
#picaxe 18X
#no_data

symbol memmin = 0            'Define the starting address for each chip
symbol memmax = 1023         'Define the ending address for each chip
symbol memaddress = w6       'I2c memory address
symbol i2cdata = b0          'Define a byte for i2c data
symbol calcdata = b1         'Define a byte for the calculated data
symbol pass = b2             'Pass count, incremented once for each pass
symbol wled = 7              'Write led
symbol rled = 6              'Read led
symbol c0led = 5             'Chip 000 led
symbol c1led = 3             'Chip 001 led
symbol eled = 2              'Error led

    pass = 0                 'Initialize the pass to zero

    do

    high wled                            'Write cycle. Turn the "write" led on
    i2cslave %10100000, i2cfast, i2cword 'Configure I2C for chip addr 000
    high c0led                           'Turn the "chip 0" led on

    for memaddress = memmin to memmax    'For all addresses
        i2cdata = memaddress + pass      'Generate some pseudo data
        writei2c memaddress, (i2cdata)   'Write data to memory
        pause 5                          'Wait for the write operation to complete
    next memaddress
    low c0led                            'Turn the "chip 0" led off
```

```
i2cslave %10100010, i2cfast, i2cword      'Configure I2C for chip addr 001
high c1led                                'Turn the "chip 1" led on

for memaddress = memmin to memmax         'For all addresses
    i2cdata = memaddress + 1 + pass       'Generate some pseudo data
    writei2c memaddress, (i2cdata)        'Write data to memory
    pause 5                               'Wait for the write operation to complete
next memaddress
low c1led                                 'Turn the "chip 1" led off

high rled       'Turn the "read" led on
pause 2000      'Pause between write and verify, with leds on
low wled        'Verify cycle. Turn the "write" led off

i2cslave %10100000, i2cfast, i2cword      'Configure I2C for chip addr 000
high c0led                                'Turn the "chip 0" led on

for memaddress = memmin to memmax         'For all addresses
    readi2c memaddress, (i2cdata)         'Read the data
    calcdata = memaddress + pass          'Regenerate data
    if i2cdata <> calcdata then           'Test if = to data that was read
        high eled                         'Data read is not expected value
        pause 10
        low eled
    endif
next memaddress

low c0led                                         'Turn the "chip 0" led off

i2cslave %10100010, i2cfast, i2cword              'Configure I2C for chip addr 001
high c1led                                        'Turn the "chip 1" led on

for memaddress = memmin to memmax         'For all addresses
    readi2c memaddress, (i2cdata)         'Read the data
    calcdata = memaddress + 1 + pass      'Regenerate data
    if i2cdata <> calcdata then           'Test if = to data that was read
        high eled                         'Data read is not expected value
        pause 10
        low eled
    endif
next memaddress

low c1led       'Turn the "chip 1" led off
low rled        'Turn the "read" led off
inc pass        'Increment the pass count

loop
```

I2C I/O Expansion

An MCP23017 I/O expander chip is used to demonstrate the use of the I2C bus for input/output expansion. The MCP23017 chip has the low-order three bits of the chip address connected to pins, which can be used to provide eight unique addresses. In this experiment, one chip is connected at address 000. This is the low-order three bits of the address, which will be combined with the high-order four bits of the address that are configured at the time of manufacture. The overall address of the chip will be 0100 000x.

The MCP23017 has two 8-bit ports that can be configured as inputs or outputs, or as a single 16-bit port. In this experiment, the 8-bit mode is used and switches are attached to port A, which is configured for input with pull-ups enabled and state inverted. LEDs are connected to port B, which is configured for output. The software initializes the ports and copies the state of the input port to the output port. When a switch is pressed, the corresponding LED will light.

The circuit is shown in Fig. 7.50

The code to configure the MCP23017 and display the switch states on LEDs is:

'I2C I/O expansion using MCP23017 - Test routine

```
#picaxe 18X
#freq m4
#gosubs 256
#no_data
```

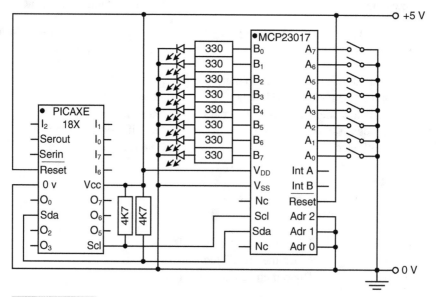

Figure 7.50 I2C I/O expansion using MCP23017.

```
symbol MCP23017 = %01000000    'Address code for MCP23017, address 000
symbol IOCON0 = $0A            'I/O control register address for bank = 0
                                  (poweron default)
symbol GPIOA = $09             'GP I/O port register
symbol GPPU = $06              'GP port pull-ups
symbol IPOL = $01              'Input port polarity
symbol GPIOB = $19             'Port B data register in eight-bit mode
symbol IODIRB = $10            'Port B direction register in eight-bit mode
symbol BANK1 = %10000000       'Bank bit for eight-bit mode

symbol i2c_data = b0           'Working register for I2C data

    i2cwrite IOCON0, (BANK1)   'Set the Bank bit in IOCON for eight-bit mode
    i2cwrite IODIRB, (0)       'Set port B to output
    i2cwrite GPPU, ($FF)       'Set port A pull-ups on
    i2cwrite IPOL, ($FF)       'Set port A polarity

    do                         'Read byte from port A and copy it to port B
        i2cread GPIOA, (i2c_data)    'Read port A
        i2cwrite GPIOB, (i2c_data)   'Write the data to port B
    loop
```

I2C Clock/Calendar

The DS1307 is a real-time clock and calendar in a single chip, which has an I2C interface, 64 bytes of RAM, and uses a 32.768 KHz crystal or external oscillator for time keeping. Seconds, minutes, hours, day of week, day of month, and year are stored in BCD format in internal registers that take up the first seven bytes of RAM. The eighth byte of RAM is taken up by a register that controls the output and frequency of a square-wave pulse output. The remaining 56 bytes of RAM are available for use as storage.

The DS1307 can be configured for 12 or 24 hour mode and the calendar is internally programmed to compensate for the number of days in a month, including leap years. There is provision for an external backup battery, which will maintain the timekeeping circuits and RAM contents if power is removed from the chip. Accuracy of the timekeeping circuits is determined by the accuracy of the crystal or external oscillator.

The layout of the DS1307 registers can be found in the data sheet and is reproduced below for convenience.

When the DS1307 is powered on, the content of the registers are indeterminate and must be initialized for proper operation.

The experiment describes a clock/calendar that uses an AXE033 LCD module fitted with a DS1307 chip and, optionally, a lithium backup battery. The AXE033 has an

DS 1307 Register Address	Use	Bit 7	Bit 6	Bit 5	Bit 4	Bit 3	Bit 2	Bit 1	Bit 0
0	Seconds	Clk halt	Seconds, 10's digit			Seconds, unit digit			
1	Minutes	0	Minutes, 10's digit			Minutes, units digit			
2	Hours	0	1 = 12 hour 0 = 24 hour	0 = AM 1 = PM Hours 10's digit	Hours 10's digit	Hours, units digit			
3	Day of week	0	0	0	0	0	Day of week (1–7)		
4	Day of month	0	0	Day of month, 10's digit		Day of month, units digit			
5	Month	0	0	0	Month 10's digit	Month, units digit			
6	Year	Year, 10's digit				Year, units digit			
7	Control	OUT	0	0	SQWE	0	0	RS1	RS0
8–63	RAM	Available for user data							

inbuilt clock mode; however, in this experiment program code is used to display and set the time and date. The circuit is shown in Fig. 7.51.

The code for the clock is:

```
'Picaxe-18X LCD I2C clock.
'Uses AXE033 in I2C mode (Jumper 1 fitted) with DS1307 fitted
#picaxe 18x
#no_data

symbol dateseparator = "–"
symbol timeseparator = ":"
```

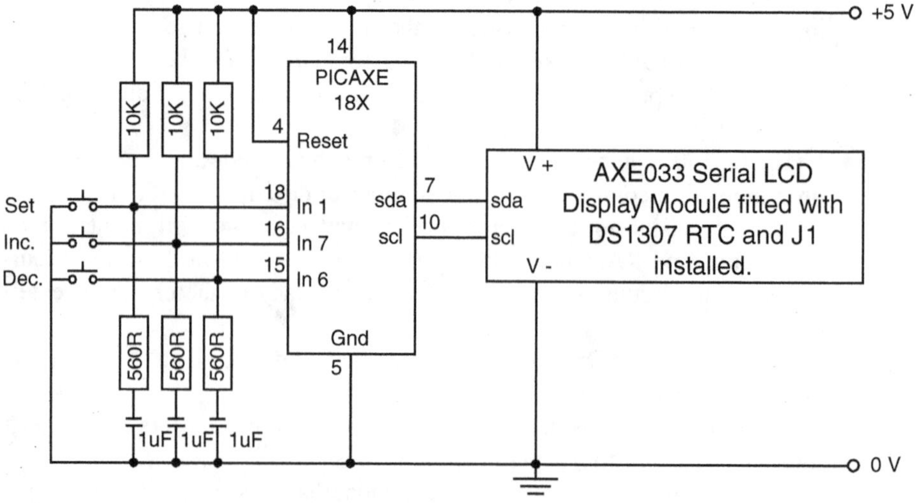

Figure 7.51 LCD clock.

```
symbol down = 0
symbol up = 1
symbol AM = 0
symbol hmode = bit6
symbol ampm = bit5

                              'LCD copmmand constants
symbol lcdclear = $01         'Clear display
symbol lcdline1 = $80         'Move to line 1 char 0
symbol lcdline2 = $C0         'Move to line 2 char 0
symbol lcdcsroff = $0C        'Cursor off
symbol lcdcsron = $0E         'Cursor on
symbol lcdcsrblink = $0F      'Cursor on blinking
symbol lcdcsrdleft = $04      'Cursor moves left
symbol lcdcsrdright = $06     'Cursor moves right

symbol setbutton = pin1
symbol incbutton = pin7
symbol decbutton = pin6

symbol work = b0              'Temporary working register
symbol bcdnum = b1            'Used for BCD arithmetic
symbol bcdmin = b2            'Used for BCD arithmetic
symbol bcdmax = b3            'Used for BCD arithmetic

symbol char1 = b4             'Temporary storage for lcd character
symbol char2 = b5             'Temporary storage for lcd character
symbol char3 = b6             'Temporary storage for lcd character

symbol seconds = b7           'DS1307 seconds
symbol mins = b8              'DS1307 minutes
symbol hour = b9              'DS1307 hour
symbol dow = b10              'DS1307 day of week
symbol dom = b11              'DS1307 day of month
symbol month = b12            'DS1307 month
symbol year = b13             'DS1307 year
```

'Operating instructions
'At turn-on, the time and date are displayed.
'Press Set button to set time and date.
'Press Inc button to increment by one.
'Press Dec button to decrement by one.
'Pressing set button cycles through day of week, day of month, month, year, am/pm,
    hour, minute, second.
If Set, Inc, or Dec buttons are held down, the program will wait until they are released.
'At first turn-on, the clock may not operate until time and date have been set.
'12 hour mode is used

```
pause 500                              'Wait for AXE033 to initialise
i2cslave $C6,i2cslow,i2cbyte           'Initialize I2C for AXE033
writei2c 0, ($FE, lcdclear, $FF)       'Clear display
pause 30                               'Wait for command to complete

do                                     'Main loop
   if setbutton = down then            'Test the time-set button
      gosub settime                    'if it's pressed then set the time and date
   else
      gosub displaytime                'if it's not pressed then display the time and date
   endif
loop
```

```
displaytime:
   i2cslave $D0, i2cslow, i2cbyte                       'Initialize I2C for DS1307
   readi2c 0, (seconds, mins, hour, dow, dom, month, year) 'Read time and date
   i2cslave $C6,i2cslow,i2cbyte                         'Initialise I2C for AXE033
   gosub writedow                      'Write "Day of week"
   gosub writedom                      'Write "Day of month"
   gosub writemonth                    'Write "Month"
   gosub writeyear                     'Write "Year"
   gosub writehour                     'Write Hours'
   gosub writeminute                   'Write "Minutes"
   gosub writesecond                   'Write "Seconds"
   return
```

```
settime:
   i2cslave $D0, i2cslow, i2cbyte                       'Initialize I2C for DS1307
   readi2c 0, (seconds, mins, hour, dow, dom, month, year) 'Read time and date
   gosub displaytime

   gosub setdow                        'Set "day of week"
   gosub setdom                        'Set "Day of month"
   gosub setmonth                      Write '"Month"'
   gosub setyear                       'Write "Year"
   gosub sethour                       'Write Hours'
   gosub setminute                     'Write "Minutes"
   gosub setsecond                     'Write "Seconds"

   do while setbutton = down loop      'Wait for button to be released
   i2cslave $D0, i2cslow, i2cbyte      'Initialize I2C for DS1307
   writei2c 0, (seconds, mins, hour, dow, dom, month, year) 'Write time and date

   i2cslave $C6,i2cslow,i2cbyte                         'Initialize I2C for AXE033
   writei2c 0, ($FE, lcdline1, $FE, lcdcsroff, $FF)   'Move to line 1 char 0, cursor off
   return
```

```
'WRITE TIME-------------------------------------------------------------------
writedow:           'Write day of week to AXE033 in I2C mode
   writei2c 0, ($FE, lcdline1, $FE, lcdcsroff, $FF)   'Move to line 1, ch 0, cursor off
   lookup dow, ("", "S", "M", "T", "W", "T", "F", "S"), char1 '1st character
   lookup dow, ("", "U", "O", "U", "E", "H", "R", "A"), char2 '2nd character
   lookup dow, ("", "N", "N", "E", "D", "U", "I", "T"), char3 '3rd character
   writei2c 0, (char1, char2, char3, " ", $FF)
   pause 10
   return

writedom:                            'Write day of month to AXE033 in I2C mode
   work = lcdline1 + 4               'Calculate line + char position
   writei2c 0, ($FE, work, $FE, lcdcsroff, $FF)   'Move to line 1 char 4, cursor off
   char1 = dom & $30 / 16 + $30      'Isolate/align ms digit; convert to ASCII
   char2 = dom & $0F + $30           'Isolate ls digit; convert to ASCII
   writei2c 0, (char1, char2, dateseparator, $FF)
   pause 10
   return

writemonth:                          'Write month to AXE033 in I2C mode
   work = lcdline1 + 7               'Calculate line + char position
   writei2c 0, ($FE, work, $FE, lcdcsroff, $FF)      'Move to line 1 char 7, cursor off
   char1 = month & $10 / 16 + $30    'Isolate/align ms digit; convert to ASCII
   char2 = month & $0F + $30         'Isolate ls digit; convert to ASCII
   writei2c 0, (char1, char2, dateseparator, $FF)
   pause 10
   return

writeyear:                           'Write year to AXE033 in I2C mode
   work = lcdline1 + 10              'Calculate line + char position
   writei2c 0, ($FE, work, $FE,      'Move to line 1 char 10, cursor off
      lcdcsroff, $FF)
   char1 = year & $F0 / 16 + $30     'Isolate/align ms digit; convert to ASCII
   char2 = year & $0F + $30          'Isolate ls digit; convert to ASCII
   writei2c 0, (char1, char2, $FF)
   pause 10
   return

writehour:                      'Write hour to A XE033 in I2C mode in 12-h mode
   writei2c 0, ($FE, lcdline2, $FE, lcdcsroff, $FF)   'Move to line 2 ch 0, cursor off
   b0 = hour                         'Move hour to b0 for bit testing
      char1 = hour & $10 / 16 + $30  'Isolate 10's digit
      char2 = hour & $0F + $30       'Isolate units digit
   if ampm = AM then
      writei2c 0, ("AM ", char1, char2, timeseparator, $FF) '12h time AM
      else
```

```
        writei2c 0, ("PM ", char1, char2, timeseparator, $FF) '12-htime PM
        pause 10
        endif
      return
```

```
writeminute:                              'Write minutes to AXE033 in I2C mode
    work = lcdline2 + 6                    'Calculate line + char position
    writei2c 0, ($FE, work, $FE,          'Move to line 2 char 6, cursor off
      lcdcsroff, $FF)
    char1 = mins & $F0 / 16 + $30          'Isolate/align ms digit; convert to ASCII
    char2 = mins & $0F + $30               'Isolate ls digit; convert to ASCII
    writei2c 0, (char1, char2, timeseparator, $FF)
    pause 10
    return
```

```
writesecond:                              'Write seconds to AXE033 in I2C mode
    work = lcdline2 + 9                    'Calculate line + char position
    writei2c 0, ($FE, work, $FE,          'Move to line 2 char 9, cursor off
      lcdcsroff, $FF)
    char1 = seconds & $F0 / 16 + $30       'Isolate/align ms digit; convert to ASCII
    char2 = seconds & $0F + $30            'Isolate ls digit; convert to ASCII
    writei2c 0, (char1, char2, $FF)
    pause 10
    return
```

```
'SET TIME-----------------------------------------------------------------
setdow:                                   'Set 'day of week'
    do while setbutton = down loop        'Wait for button to be released
    bcdmax = $07: bcdmin = $01            'Set limits
    do
      writei2c 0, ($FE, lcdline1, $FE, lcdcsrblink,   'Move to line 1 ch 0, blink
        $FF)
      if incbutton = down then
        bcdnum = dow gosub bcdinc dow = bcdnum
      gosub writedow                      'Display current value
        do while incbutton = down loop    'Wait for button to be released
        endif
      if decbutton = down then
        bcdnum = dow gosub bcddec dow = bcdnum
      gosub writedow                      'Display current value
        do while decbutton = down loop    'Wait for button to be released
        endif
      if setbutton = down then exit
    loop
    return
```

```
setdom:                                    'Set 'day of month'
    do while setbutton = down loop         'Wait for button to be released
    bcdmax = $31: bcdmin = $01             'Set limits. Assume 31 for all months
    do
        work = lcdline1 + 4                        'Calculate line + char position
        writei2c 0, ($FE, work, $FE, lcdcsrblink, $FF)    'Move to line 1 ch 4, blink
        bcdnum = dom
        if incbutton = down then
            bcdnum = dom gosub bcdinc dom = bcdnum
            gosub writedom                 'Display current value
            do while incbutton = down loop 'Wait for button to be released
            endif
        if decbutton = down then
            bcdnum = dom gosub bcddec dom = bcdnum
            gosub writedom                 'Display current value
            do while decbutton = down loop 'Wait for button to be released
            endif
        if setbutton = down then exit
    loop
    return

setmonth:                                  'Set 'month'
    do while setbutton = down loop         'Wait for button to be released
    bcdmax = $12: bcdmin = $01             'Set limits.
    do
        work = lcdline1 + 7                        'Calculate line + char position
        writei2c 0, ($FE, work, $FE, lcdcsrblink, $FF)'Move to line 1 ch 7, cursor blink
        if incbutton = down then
            bcdnum = month gosub bcdinc month = bcdnum
            gosub writemonth               'Display current value
            do while incbutton = down loop 'Wait for button to be released
            endif
        if decbutton = down then
            bcdnum = month gosub bcddec month = bcdnum
            gosub writemonth               'Display current value
            do while decbutton = down loop 'Wait for button to be released
            endif
        if setbutton = down then exit
    loop
    return

setyear:                                   'Set 'year'
    do while setbutton = down loop         'Wait for button to be released
    bcdmax = $99: bcdmin = $01             'Set limits.
    do
```

```
       work = lcdline1 + 10                  'Calculate line + char position
       writei2c 0, ($FE, work, $FE, lcdcsrblink, $FF)   'Move to line 1 ch 10, blink
       if incbutton = down then
           bcdnum = year gosub bcdinc year = bcdnum
           gosub writeyear                   'Display current value
           do while incbutton = down loop    'Wait for button to be released
           endif
       if decbutton = down then
           bcdnum = year gosub bcddec year = bcdnum
           gosub writeyear                   'Display current value
           do while decbutton = down loop    'Wait for button to be released
           endif
       if setbutton = down then exit
   loop
   return


sethour:                       'Set AM/PM indicator, then set hour
   hour = hour | $40           'Set 12-h mode
   gosub setampm               'Set AM/PM
   gosub sethr                 'Set hours
   return


setampm:                                          'Set AM / PM for 12 hour mode
   do while setbutton = down loop                 'Wait for button to be released
   bcdmax = $01: bcdmin = $00                      'Set limits.
   do
       writei2c 0, ($FE, lcdline2, $FE, lcdcsrblink, $FF)   'Move to line 2 char 0, blink
       if incbutton = down then
           bcdnum = hour & $20 / 32               'Isolate and position AM / PM bit
           gosub bcdinc
           bcdnum = bcdnum * 32                    'Reposition AM / PM bit
           hour = hour & $DF | bcdnum              'Insert the AM/PM bit
           gosub writehour                         'Display current value
           do while incbutton = down loop          'Wait for button to be released
           endif
       if decbutton = down then
           bcdnum = hour & $20 / 32               'Isolate and position AM / PM bit
           gosub bcddec
           bcdnum = bcdnum * 32                    'Reposition AM / PM bit
           hour = hour & $DF | bcdnum              'Insert the AM/PM bit
           gosub writehour                         'Display current value
           do while decbutton = down loop          'Wait for button to be released
           endif
       if setbutton = down then exit
   loop
   return
```

```
sethr:                                      'Set 'hour'
    do while setbutton = down loop          'Wait for button to be released
    bcdmax = $12 bcdmin = $01
    do
        work = lcdline2 + 3                 'Calculate line + char position
        writei2c 0, ($FE, work, $FE, lcdcsrblink, $FF)   'Move to line 2 ch 3, blink
        if incbutton = down then
            bcdnum = hour & $1F gosub bcdinc hour = hour & $E0 | bcdnum
            gosub writehour                 'Display current value
            do while incbutton = down loop  'Wait for button to be released
            endif
        if decbutton = down then
            bcdnum = hour & $1F gosub bcddec hour = hour & $E0 | bcdnum
            gosub writehour                 'Display current value
            do while decbutton = down loop  'Wait for button to be released
            endif
        if setbutton = down then exit
    loop
    return

setminute:                                  'Set 'minute'
    do while setbutton = down loop          'Wait for button to be released
    bcdmax = $59: bcdmin = $00              'Set limits.
    do
        work = lcdline2 + 6                 'Calculate line + char position
        writei2c 0, ($FE, work, $FE, lcdcsrblink,    'Move to line 2 ch 6, cursor blink
            $FF)
        if incbutton = down then
            bcdnum = mins gosub bcdinc mins = bcdnum
            gosub writeminute               'Display current value
            do while incbutton = down loop  'Wait for button to be released
            endif
        if decbutton = down then
            bcdnum = mins gosub bcddec mins = bcdnum
            gosub writeminute               'Display current value
            do while decbutton = down loop  'Wait for button to be released
            endif
        if setbutton = down then exit
    loop
    return

setsecond:                                  'Set 'second'
    do while setbutton = down loop          'Wait for button to be released
    bcdmax = $59: bcdmin = $00              'Set limits.
    do
        work = lcdline2 + 9                 'Calculate line + char position
        writei2c 0, ($FE, work, $FE, lcdcsrblink, $FF) 'Move to line 2 ch 9, cursor blink
```

```
      if incbutton = down then
         bcdnum = seconds gosub bcdinc seconds = bcdnum
         gosub writesecond                    'Display current value
         do while incbutton = down loop       'Wait for button to be released
         endif
      if decbutton = down then
         bcdnum = seconds gosub bcddec seconds = bcdnum
         gosub writesecond                    'Display current value
         do while decbutton = down loop       'Wait for button to be released
         endif
      if setbutton = down then exit
   loop
   return


bcdinc:   'Increment by 1, BCD numbers that are stored 2 per byte and maintain limits
   bcdnum = bcdnum + 1         'Increment BCD number
   work = bcdnum & $0F         'Isolate least significant digit
   if work = 10 then           'Test for overflow
      bcdnum = bcdnum + 6      'Overflow occurred - set ls digit = 0 and inc ms digit
   endif

   if bcdnum > bcdmax then     'Test for maximum limit
      bcdnum = bcdmin          'Over limit - reset to minimum value
   endif
   return


bcddec:   'Decrement by 1, BCD numbers that are stored 2 per byte and maintain
             limits
   if bcdnum = bcdmin then     'Test for minimum limit
      bcdnum = bcdmax          'If at minimum value, set to maximum
      else
      bcdnum = bcdnum - 1      'Decrement bcdnum
      work = bcdnum & $0F      'Isolate least significant digit 'and
      if work >= $0F then      'test for underflow and
      bcdnum = bcdnum - 6      'correct for underflow
      endif
   endif
   return
```

SPI Memory Expansion

Two 25LC256 SPIEEPROM chips are used to demonstrate the use of the SPI bus for memory expansion and the use of more than one chip. The 25LC256 is addressed by chip-select pins that are controlled by the PICAXE; two chips are used in this experiment.

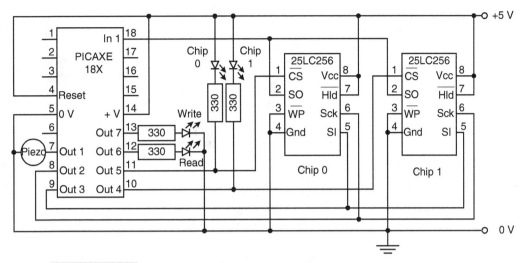

Figure 7.52 The SPI memory expansion circuit.

SPI memory chips are like a miniprocessor and instructions must be issued to them in order to obtain the desired result. Some operations, such as writes, can continue on in the background after an instruction has been sent and it is, therefore, necessary to test a chip to see if it is busy before issuing any further instructions. In addition, the chips are write-disabled at power on and after a write instruction and, therefore, must be write-enabled prior to issuing a write instruction.

The example code uses bit-banging to control the SPI bus. In practice, the spi_so and spi_si subroutines can be easily replaced with **shiftout, shiftin,** or the hardware SPI commands.

```
'SPI Memory expansion using two 25LC256 - Test routine
#picaxe 18X
#freq m4
#gosubs 256
#no_data


symbol CS0 = 5            'Chip 0 enable
symbol CS1 = 4            'Chip 1 enable
symbol SCK = 2            'SPI serial clock
symbol MOSI = 3           'SPI serial out
symbol MISO = pin1        'SPI serial in


symbol maxpasses = 255    'Number of passes ( 0 – 255).
symbol maxchips = 1       'Number of chips - 1 (first chip = 0)
'symbol maxaddress = 32767 'Chip size in bytes - 1 (25LC256 = 32,767).
symbol maxaddress = 63    'Limit the number of bytes to reduce the time taken
```

```
symbol c_error = 1              'Piezo sounder
symbol c_write = 7              'Write LED
symbol c_read = 6               'Read LED

symbol SPIBUSY = bit0           'SPI "busy" bit
symbol YES = 1

                                '25C256 Opcodes
symbol WRSR = %00000001         'Write status register
symbol SWRITE = %00000010       'Write data
symbol SREAD = %00000011        'Read data
symbol WRDI = %00000100         'Write disable
symbol RDSR = %00000101         'Read status register
symbol WREN = %00000110         'Write enable

symbol spi_wdata = b1           'Data to be written
symbol spi_rdata = b2           'Data that was read
symbol spi_chip = b3            'SPI chip ID
symbol spi_status = b4          'Status byte
symbol spir_ctr = b5            'Counter used in SPI routines
symbol spir_outbyte = b6        'SPI data byte for write
symbol pcount = b7              'Pass counter
symbol spi_addr = w5            'SPI address, must be a word with two bytes mapped to it
symbol spi_addrlo = b10         'Address low byte, maps to spi_addr high byte
symbol spi_addrhi = b11         'Address high byte, maps to spi_addr low byte

sound c_error, (10, 10)         'Beep to indicate start of program
                                'Initialize SPI memory chips
                                'All chips will be in selected state at startup because CS is
                                'active low. So it's necessary to deselect them all.
for spi_chip = 0 to 1           'For all spi chips
   gosub spi_dsel               'Deselect the current chip
next spi_chip


do

   for pcount = 0 to maxpasses              'Make several passes through
                                            '   the chips
      for spi_chip = 0 to maxchips          'For each chip
         high c_write                       'Turn write LED on
         for spi_addr = 0 to maxaddress     'For each address
            spi_wdata = pcount + spi_chip + spi_addr   'Generate pseudo data
            gosub spi_wrb                   'Write a byte
         next spi_addr                      'Next address
         low c_write                        'Turn write led off
```

```
            high c_read                                'Turn read LED on
        for spi_addr = 0 to maxaddress                 'For each address
            gosub spi_rdb                              'Read a byte
            spir_outbyte = pcount + spi_chip + spi_addr 'Regenerate the data
            if spi_rdata <> spir_outbyte then          'Compare data
                sound c_error, (10, 10)                'Beep to indicate error
            endif
        next spi_addr                                  'Next address
            low c_read                                 'turn read LED off

    next spi_chip      'Next chip

  next pcount      'Next pass

loop     'Start again and repeat forever
```

'***

'SPI subroutines for memory chips with 16-bit addresses

```
'   spi_busy       Wait while chip is busy
'   spi_wrst       Write the status byte
'   spi_rds        Read status byte
'   spi_wrb        Write a single byte
'   spi_rdb        Read a single byte
'   spi_wren       Write-enable the chip
'   spi_sel        Select the chip
'   spi_dsel       Deselect the chip
'   spi_so         Write a byte of data from spi_wdata
'   spi_si         Read a byte of data into spi_rdata
```

'Variables used
'
```
'   spi_chip       Contains the chip ID (0 - n), where n = number of chips - 1
'   spi_addr       The 25LC256 address. Must be a word and have hi and lo bytes
'   spi_wdata      The data to be written
'   spi_data       The data that was read
'   spi_status     The status byte to be written
'   spir_ctr       Counter for the bits to be transmitted,
'   spir_outbyte   A work area for the data byte that will be written.
```

'On entry
'
```
'   spi_chip contains the chip ID (0 - n)
'   spi_addrhi contains the high byte of the address to be read/written
'   spi_addrlo contains the low byte of the address to be read/written
'   spi_wdata contains the data byte to be written
'   spi_status contains the status byte to be sent
'
```

```
'On exit
'
'   spi_chip is unchanged
'   spi_addr is unchanged
'   spi_wdata is unchanged
'   spi_rdata contains the byte which was read, unchanged for write operation
'   spi_status is unchanged

spi_wrb:                        'Write a single byte from "spi_wdata"
                                'Requires the chip number in spi_chip
                                'Requires the data byte to be written in spi_wdata
                                'Requires address in spi_addr
    gosub spi_busy              'Wait if busy
    gosub spi_wren              'Write enable
    gosub spi_sel               'Select chip
    spir_outbyte = SWRITE       'Set opcode
    gosub spi_so                'Write opcode
    spir_outbyte = spi_addrhi   'Set address high
    gosub spi_so                'Write address high
    spir_outbyte = spi_addrlo   'Set address low
    gosub spi_so                'Write address low
    spir_outbyte = spi_wdata    'Set data
    gosub spi_so                'Write data
    gosub spi_dsel              'Deselect chip
    return

spi_rdb:                        'Read a single byte into "spi_rdata"
                                'Requires the chip number in spi_chip
                                'Requires address in spi_addr
                                'On exit data will be read into spi_rdata
    gosub spi_busy              'Wait if busy
    gosub spi_sel               'Select chip
    spir_outbyte = SREAD        'Set opcode
    gosub spi_so                'Write opcode
    spir_outbyte = spi_addrhi   'Set address high
    gosub spi_so                'Write address high
    spir_outbyte = spi_addrlo   'Set address low
    gosub spi_so                'Write address low
    gosub spi_si                'Read data
    spi_rdata = b0              'Data was read into b0, put it into spi_data
    gosub spi_dsel              'Deselect chip
    return

spi_busy:                       'Test if device is busy
                                'Requires the chip number in spi_chip
    gosub spi_sel               'Select chip
```

```
spir_outbyte = RDSR              'Set opcode
gosub spi_so                     'Write the opcode
gosub spi_si                     'Read the status into b0
gosub spi_dsel                   'De-select chip
if SPIBUSY = YES then spi_busy   'Status is still in b0, if busy bit is set then wait
return

spi_wren:                        'Write enable
                                 'Busy status must already have been tested prior to entry
                                     'Requires the chip number in spi_chip

gosub spi_sel                    'Select chip
spir_outbyte = WREN              'Set opcode
gosub spi_so                     'Write opcode
gosub spi_dsel                   'Deselect chip
return

spi_rdst:                        'Read status into "spi_rdata"
                                 'No need to wait if busy, status can be read anytime
                                 'Requires the chip number in spi_chip

gosub spi_sel                    'Select chip
spir_outbyte = RDSR              'Set opcode
gosub spi_so                     'Write opcode
gosub spi_si                     'Read the status
gosub spi_dsel                   'Deselect chip
return

spi_wrst:                        'Write status register
                                 'Requires the chip number in spi_chip
                                 'Requires status value in spi_status

gosub spi_busy                   'Wait if busy
gosub spi_wren                   'Write enable
gosub spi_sel                    'Select chip
spir_outbyte = WRSR              'Set opcode
gosub spi_so                     'Write opcode
spir_outbyte = spi_status        'Status register value
gosub spi_so                     'Write status
gosub spi_dsel                   'Deselect chip
return

spi_sel:                         'Select chip
                                 'Requires chip# in spi_chip

low SCK
select spi_chip
case 0 low CS0                   'Add more cases here for more chips
case 1 low CS1
endselect
return
```

```
spi_dsel:                    'Deselect chip
                             'Requires chip# in spi_chip
   low SCK
   select spi_chip
   case 0 high CS0           'Add more cases here for more chips
   case 1 high CS1
   endselect
   return


spi_so:                      'Write a byte from spir_outbyte
   b0 = spir_outbyte         'Move output byte to b0 for bit testing and input/output
   for spir_ctr = 1 to 8     'For each of eight bits
      if bit7 = 1 then       'Next bit is 1, set MOSI high
         high MOSI
      else
         low MOSI            'Next bit is 0, set MOSI low
      endif
      high SCK               'Clock high
      b0 = b0 * 2            'Shift data left
      low SCK                'Clock low
   next spir_ctr
   return


spi_si:                      'Read a byte into spi_data
   for spir_ctr = 1 to 8     'For each of eight bits
      high SCK               'Clock high
      b0 = b0 * 2 | MISO     'Shift data left and read MISO into ls bit
      low SCK                'Clock low
   next spir_ctr
   spi_rdata = b0            'Byte read is in b0, put it in spi_rdata
   return
```

SPI I/O Expansion

An MCP23S17 I/O expander chip is used to demonstrate the use of the SPI bus for input/output expansion. The MCP23S17 chip has the low-order three bits of the chip address connected to pins, which can be used to provide eight unique addresses and also has a chip-select pin that is controlled by the PICAXE. In this experiment, the external address pins are disabled and the chip-select pin is used to select the chip.

The MCP23S17 has two 8-bit ports that can be configured as inputs or outputs, or as a single 16-bit port. In this experiment, 8-bit mode is used and switches are attached to port A, which is configured for input with pull-ups enabled and state inverted. LEDs are connected to port B, which is configured for output. The software initializes the

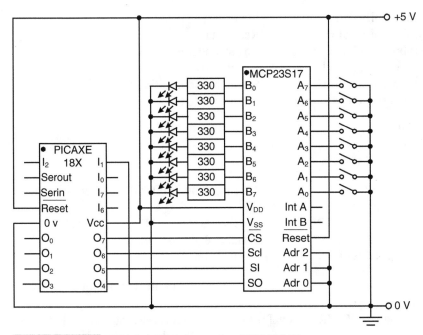

Figure 7.53 SPI I/O expansion using MCP23S17.

ports and copies the state of the input port to the output port. When a switch is pressed, the corresponding LED will light. The circuit is shown in Fig. 7.53

The example code uses bit-banging to control the SPI bus. In practice, the spi_so and spi_si subroutines can be replaced with **shiftout, shiftin,** or hardware SPI commands.

```
'SPI I/O expansion using MCP23S17 – Test routine

#picaxe 18X
#freq m4
#gosubs 256
#no_data

symbol CS0 = 7            'Chip enable
symbol SCK = 6            'SPI serial clock
symbol MOSI = 5           'SPI serial out
symbol MISO = pin1        'SPI serial in

symbol MCP23S17_OPCODE = %01000000   'Op code for MCP23S17, address 000
symbol IOCON0 = $0A  'I/O control register address for bank = 0 (poweron default)
symbol IODIR = $00        'I/O direction
symbol GPIO = $09         'GP I/O port register
symbol GPPU = $06         'GP port pull-ups
symbol IPOL = $01         'Input port polarity
symbol PORTB_OFFS = $10   'Port B offset (add this to base address for port B)
symbol BANK1 = %10000000  'Bank bit for eight-bit mode
```

```
        symbol spi_wdata = b1        'Write data
        symbol spi_rdata = b2        'Read data
        symbol spi_addr = b3         'Register address
        symbol spi_chip = b4         'SPI chip ID
        symbol spir_ctr = b5         'Counter used in SPI routines
        symbol spir_outbyte = b6     'SPI data byte for write

    initialize:
        spi_chip = 0        'Chip 0
        gosub spi_dsel


                                'Set the Bank bit in IOCON for eight-bit mode
        spi_chip = 0            'Chip 0
        spi_addr = IOCON0       'Set register address = IOCON in bank 0
        spi_wdata = BANK1       'Set data = eight-bit mode
        gosub spi_write         'Write the data


                                        'Set port B to output
        spi_chip = 0                    'Chip 0
        spi_addr = IODIR + PORTB_OFFS   'Set register address = IODIRB
        spi_wdata = $00                 'Set data = 0 (all pins output)
        gosub spi_write                 'Write the data


                        'Set port A pull-ups on
        spi_chip = 0    'Chip 0
        spi_addr = GPPU 'Set register address
        spi_wdata = $FF 'Set data = 1's (all pull-ups active)
        gosub spi_write 'Write the data


                        'Set port A polarity
        spi_chip = 0    'Chip 0
        spi_addr = IPOL 'Set register address
        spi_wdata = $FF 'Set data = 1's (all polarity inverted)
        gosub spi_write 'Write the data

    do

                                'Read byte from port A and copy it to port B
        spi_chip = 0            'Chip 0
        spi_addr = GPIO         'Set register address = Port A
        gosub spi_read          'Read the data

        spi_addr = GPIO + PORTB_OFFS    'Set register address = Port B
        spi_wdata = spi_rdata           'Set data = input data
        gosub spi_write                 'Write the data

    loop
```

```
'***************************************************************
'SPI subroutines
'   spi_write      Write to a register
'   spi_read       Read from a register
'   spi_sel        Select the chip
'   spi_dsel       Deselect the chip
'   spi_so         Write a byte
'   spi_si         Read a byte

'On entry
'
'   spi_chip contains the chip ID (0 - n)
'   spi_addr contains the address of the register
'   spi_wdata contains the data to be written
'
'On exit
'
'   spi_chip remains unchanged
'   spi_addr remains unchanged
'   spi_outbyte remains unchanged
'   spi_rdata contains the data that was read, remains unchanged for write
'   spi_wdata remains unchanged

spi_write:                                  'Write to a register
                                            'Chip ID in spi_chip
                                            'Requires register address in spi_addr
                                            'Register data in spi_wdata
    gosub spi_sel                           'Select chip
    spir_outbyte = MCP23S17_OPCODE          'Op code for write
    gosub spi_so                            'Write op code
    spir_outbyte = spi_addr                 'Register address
    gosub spi_so                            'Write register address
    spir_outbyte = spi_wdata                'Data
    gosub spi_so                            'Write data
    gosub spi_dsel                          'Deselect the chip
    return

spi_read:                                   'Read from a register
                                            'Chip ID in spi_chip
                                            'Requires register address in spi_addr
                                            'Register data will be returned in
                                            '  spi_rdata
    gosub spi_sel                           'Select the chip
    spir_outbyte = MCP23S17_OPCODE | $01    'Op code for read
    gosub spi_so                            'Write op code
```

```
      spir_outbyte = spi_addr  'Register address
      gosub spi_so             'Write register address
      gosub spi_si             'Read data
      gosub spi_dsel           'Deselect the chip
      return


spi_sel:                       'Select chip
                               'Requires chip# in spi_chip

   low SCK
   select spi_chip
      case 0 low CS0           'Add extra case clauses for additional chips
   endselect
   return


spi_dsel:                      'Deselect chip
                               'Requires chip# in spi_chip

   select spi_chip
      case 0 high CS0          'Add extra case clauses for additional chips
   endselect
   return


spi_so:                        'Write a byte from spir_outbyte
   b0 = spir_outbyte           'Move the output byte to b0 for bit testing
   for spir_ctr = 1 to 8       'For each of eight bits
      if bit7 = 1 then         'Test the bit
         high MOSI             'Next bit is 1, set MOSI high
         else
         low MOSI              'Next bit is 0, set MOSI low
      endif
      high SCK                 'Clock high
      b0 = b0 * 2              'Shift data left
      low SCK                  'Clock low
   next spir_ctr
   return

spi_si:                        'Read a byte into spi_data
   for spir_ctr = 1 to 8       'For each of eight bits
      high SCK                 'Clock high
      b0 = b0 * 2 | MISO       'Shift data left and read MISO into ls bit
      low SCK                  'Clock low
   next spir_ctr
   spi_rdata = b0              'Byte read is in b0, put it in spi_rdata
   return
```

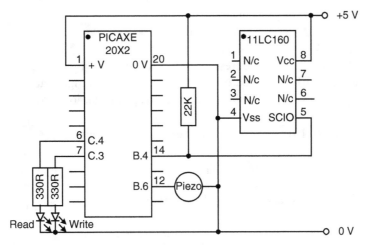

Figure 7.54   UNI/O memory expansion.

UNI/O^{TM} Memory Expansion

An 11LC160 chip is used to demonstrate the use of the UNI/O bus for memory expansion. The 11LC160 is addressed by an internal address that is fixed at the time of manufacture. The circuit for the UNI/O experiment is shown in Fig. 7.54.

UNI/O memory chips are like a miniprocessor and instructions must be issued to them in order to obtain the desired result. Some operations, such as writes, can continue on in the background after an instruction has been sent. It is, therefore, necessary to test a chip to see if it is busy before issuing any further instructions. In addition, the chips are write-disabled at power on and after a write instruction and, therefore, must be write-enabled prior to issuing a write instruction.

The code to access the 11LC160 is:

```
'UNI/O Memory expansion using a single 11LC160 - Test routine
#picaxe 20X2
#no_data

symbol maxpasses = 255        'Number of passes
symbol maxaddress = 2047      'Chip size in bytes - 1 (11LC160 = 2048 bytes).

symbol SCIO = B.4        'Serial input/output pin
symbol c_error = B.6     'Piezo sounder
symbol c_write = C.3     'Write LED
symbol c_read = C.4      'Read LED

symbol UNIBUSY = bit0        'UNI/O "busy" bit
symbol UCA = %10100000       'Chip address for 25LC160
symbol YES = 1
```

```
symbol uni_wdata = b1       'Data to be written
symbol uni_rdata = b2       'Data that was read
symbol pcount = b3          'Pass counter
symbol uni_addr = w5        'Address, must be a word with 2 bytes mapped to it
symbol uni_addrlo = b10     'Address low byte, maps to addr high byte
symbol uni_addrhi = b11     'Address high byte, maps to addr low byte

    sound c_error, (10, 10)     'Beep to indicate start of program

    do                          'Main loop

    for pcount = 0 to maxpasses     'Make several passes

        high c_write                            'Turn write LED on
        for uni_addr = 0 to maxaddress          'For each address
        uni_wdata = pcount + uni_addr           'Generate pseudo data
        gosub uni_busy                          'Wait if busy
        uniout SCIO, UCA, UNI_WREN              'Write enable
        uniout SCIO, UCA, UNI_Write, uni_addrhi, uni_addrlo, (uni_wdata)
        next uni_addr                           'Next address
        low c_write                             'Turn write LED off

        high c_read                             'Turn read LED on
        for uni_addr = 0 to maxaddress          'For each address
        gosub uni_busy                          'Wait if busy
        uniin SCIO, UCA, UNI_READ, uni_addrhi, uni_addrlo, (uni_rdata)
        uni_wdata = pcount + uni_addr           'Re-generate the data
        if uni_rdata <> uni_wdata then          'Compare data
            sound c_error, (10, 10)             'Beep to indicate error
            endif
        next uni_addr                           'Next address
        low c_read                              'Turn read LED off

    next pcount     'Next pass

    loop

uni_busy:                                       'Test if device is busy
    uniin SCIO, UCA, UNI_RDSR, (b0)             'Read status register
    if UNIBUSY = YES then uni_busy              'If WIP (busy) is set then wait
    return
```

8

PICAXE M2 SUPPLEMENT

Introduction

While this book was being typeset, the PICAXE-18M2 was released and the PICAXE-14M2 and PICAXE-20M2 were announced. The reduced-price M2 series represents a revolution in PICAXE design and architecture by adding new features, including:

Multitasking with up to four start points in a program

Program memory for up to 1800 lines of BASIC code

28-byte variables

256 bytes RAM

Clock speed up to 32 MHz

Configurable I/O pins

Touch sensing

Digital to analog conversion

Elapsed time counter

1.8 to 5.5V operation

New commands for the M2 series chips include:

daclevel

dacsetup

disabletime

enabletime

fvrsetup

readdac

readdac10

restart

resume

srlatch

srset

srreset

suspend

touch

touch16

and the

#simtask directive

Powering the PICAXE M2 Series Chips

M2 series chips will run with voltage supplies of 1.8 to 5.5 V and all of the power supply arrangements shown in Chapter 2 can be used. In addition, the M2 series chips can be powered from 2 × AA alkaline cells and 2 × AA rechargeable cells.

The following constrains should be observed when running at low voltages:

■ M2 brownout detection operates at 1.9 V and, therefore, must be turned off when using 1.8-V supplies. See **disablebod** and **enablebod** commands.

TABLE 8.1   SUMMARY OF M2 SERIES ARCHITECTURE

CHIP	PROGRAM MEMORY (APPROX. LINES OF CODE)	VARIABLES (BYTES)	EEPROM (BYTES)	I/O PINS	ANALOG INPUTS	TOUCH INPUTS	SUPPLY VOLTAGE (V)
14M2	Up to 1800	28	256	12	7	7	1.8–5.5
18M2	Up to 1800	28	Up to 256	16	10	10	1.8–5.5
20M2	Up to 1800	28	256	16	11	11	1.8–5.5

■ Voltages applied to input pins must not exceed the supply voltage by more than +0.3 V.

■ The USB programming cables should always be used when the supply voltage is 3.3 V or less.

Resetting the M2 Series Chips

The M2 series chips can be reset by removing and restoring power and by executing the **reset** command.

Downloading Programs to the M2 Series Chips

All of the program download arrangements, shown in Chapter 2, can be used with the M2 series chips. Note that USB programming cables should be used when the supply voltage is 3.3 V or less.

Clocking the M2 Series Chips

The M2 series chips have a default clock frequency of 4 MHz and may be set to any of 31 kHz, 250 kHz, 500 kHz, 1 MHz, 2 MHz, 4 MHz, 8 MHz, 16 MHz, and 32 MHz by means of the **setfreq** command.

Memory Arrangement for M2 Series Chips

In general, the memory arrangements, shown in Chapter 2, apply to the M2 series chips, although there are some constraints in the use of EEPROM (see Fig. 8.1, later).

TABLE 8.2 PICAXE M2 CLOCK FREQUENCIES

CHIP	DEFAULT FREQUENCY (MHz)	INTERNAL CLOCK FREQUENCIES (MHz)	EXTERNAL RESONATOR FREQUENCIES (MHz)
14M2	4	0.031, 0.25, 0.5, 1, 2, 4, 8, 16, 32	Not available
18M2	4	0.031, 0.25, 0.5, 1, 2, 4, 8, 16, 32	Not available
20M2	4	0.031, 0.25, 0.5, 1, 2, 4, 8, 16, 32	Not available

TABLE 8.3   SUMMARY OF M2 SERIES DATA STORAGE

CHIP	GENERAL-PURPOSE VARIABLES			SCRATCHPAD (BYTES)	STORAGE VARIABLES (BYTES)	EEPROM (BYTES)
	WORDS	BYTES	BITS			
14M2	14	28	32	Not available	256	256
18M2	14	28	32	Not available	256	Up to 256
					128 when multitasking	
20M2	14	28	32	Not available	256	256

General-Purpose Variables for the M2 Series Chips

The discussion of general-purpose variables in Chapter 2 is valid for the M2 series; the arrangement is summarized in Table 8.3.

System Variables for M2 Series Chips

The system variables, shown in Chapter 2, have been updated in the M2 series chips. In particular, the s_w0 variable has been added and the s_w7 elapsed time variable is in use.
   The updated system variables for the M2 series are:

s_w0	task	Current task during parallel processing (M2 series)
s_w1		Reserved for future use
s_w2		Reserved for future use
s_w3	timer3	timer3 value (X2 series)
s_w4	compvalue	comparator results (X2 series)
s_w5	hserptr	hardware serin pointer
s_w6	hi2clast	hardware hi2c last byte written (slave mode)
s_w7	time	elapsed time (M2 series)

Storage Variables for the M2 Series Chips

The M2 series have 256 bytes of storage variables that are accessed by the **peek** and **poke** commands and the pointer **bptr** (see Fig. 8.2, later).
   The storage variables for M2 series chips are summarized in Table 8.4; see also Fig. 8.2.

TABLE 8.4 M2 STORAGE VARIABLES		
CHIP	BYTES	ADDRESSES
14M2	256	0–255 (includes variables b0–b27 at addresses 0–27)
18M2	256	0–255 (includes variables b0–b27 at addresses 0–27)
		0–127 when multitasking.
20M2	256	0–255 (includes variables b0–b27 at addresses 0–27)

Special Function Variables

In general, the discussion of special-function variables in Chapter 2 is valid for M2 series chips, although some changes are for the M2 series chips. In particular, the **time** (s_w7) and **task** (s_w0) variables have been added.

The special-function variables for M2 series chips are summarized in Table 8.5.

EEPROM

In general, the discussion of EEPROM in Chapter 2 is valid for M2 series chips, although the EEPROM for the 18M2 series chips is shared with the high-order 256 bytes of program memory, as shown in Fig. 8.1. (The 14M2 and 20M2 have separate program memory.)

TABLE 8.5 SPECIAL-FUNCTION VARIABLES FOR M2 SERIES CHIPS			
SPECIAL-FUNCTION VARIABLE	14M2	18M2	20M2
pinsB	✓	✓	✓
outpinsB	✓	✓	✓
dirsB	✓	✓	✓
pinsC	✓	✓	✓
outpinsC	✓	✓	✓
dirsC	✓	✓	✓
Bptr	✓	✓	✓
@bptr	✓	✓	✓
@bptrinc	✓	✓	✓
@bptrdec	✓	✓	✓
Time	✓	✓	✓
Task	✓	✓	✓

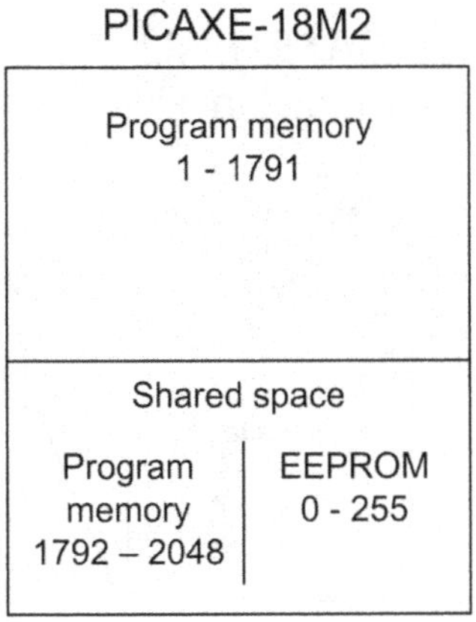

PICAXE-18M2

```
┌─────────────────────────┐
│                         │
│    Program memory       │
│       1 - 1791          │
│                         │
│                         │
│                         │
├─────────────────────────┤
│     Shared space        │
├────────────┬────────────┤
│  Program   │   EEPROM   │
│  memory    │   0 - 255  │
│ 1792 – 2048│            │
└────────────┴────────────┘
```

Figure 8.1   EEPROM arrangement for 18M2 series chips.

If a program is longer than 1791 bytes, then the subsequent bytes will take up space in the area allocated for EEPROM, thus reducing the amount of EEPROM space available to the program.

Pointers

In general, the discussion of pointers in Chapter 2 is valid for M2 series chips, although it should be noted that the byte-scratchpad pointer (bptr) is the only pointer available for the M2 series.

BYTE-SCRATCHPAD POINTER

The byte-scratchpad pointer points to addresses in the byte-scratchpad, which is made up of byte variables and storage variables, as shown later in Fig. 8.2.

For the M2 series chips, there are 256 bytes in the byte-scratchpad. The first 28 are the variables b0 to b27. However, when multitasking with the PICAXE-18M2, the high-order 128 bytes are reserved by the system, leaving 128 bytes available for use. The PICAXE-14M2 and PICAXE-20M2 have 256 bytes of byte-scratchpad available for use when multitasking. All general-purpose variables can be addressed by using the byte-scratchpad pointer.

The arrangement of the byte-scratchpad for M2 series chips is shown in Fig. 8.2

Ports

In general, the discussion of ports in Chapter 2 is valid for the M2 series chips.

M2 Series chips have two ports, B and C. Both may be used for input or output, although some bits are input and some bits are output only. M and X series chips

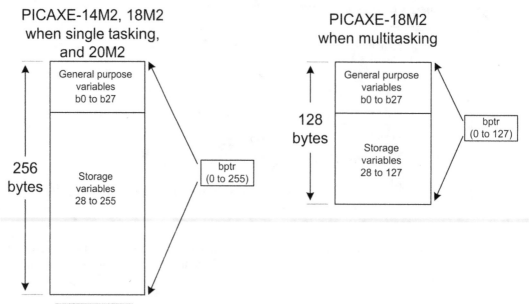

Figure 8.2 The byte-scratchpad pointer for M2 series chips.

had a dedicated input port corresponding to M2 port C and a dedicated output port corresponding to M2 port B. Port addressing compatibility with earlier chips can be achieved by configuring the M2 port direction variable (dirs) for port B to outputs and for port C to inputs. Table 8.6 shows the arrangement of digital ports for M2 series chips. Many of these ports also have other uses, such as touch-input and ADC input.

Table 8.7 shows the layout of the **pins**, **outpins**, and **dirs** variables for the PICAXE-14M2. Some pins are not implemented (see Table 8.6).

Table 8.8 shows the layout of the **pins**, **outpins**, and **dirs** variables for the PICAXE-18M2. Some pins are not implemented (see Table 8.6).

Table 8.9 shows the layout of the **pins**, **outpins**, and **dirs** variables for the PICAXE-20M2. Some pins are not implemented (see Table 8.6).

Setting the Direction of Configurable Pins

The direction of pins that can be either inputs or outputs can be set by means of the **input** and **output** commands, by a **high** or **low** command referencing the pin, or by setting the corresponding bit in the **dirs** variable.

Examples for PICAXE-14M2, 18M2, and 20M2:

TABLE 8.6 PICAXE-M2 DIGITAL PORT ASSIGNMENTS

	PICAXE-14M2		PICAXE-18M2		PICAXE-20M2	
BIT	PORT B	PORT C	PORT B	PORT C	PORT B	PORT C
7	Not available	Not available	Input B.7 Output B.7 Output 7	Input C.7 Output C.7 Input 7	Input B.7 Output B.7 Output 7	Input C.7 Output C.7 Input 7
6	Not available	Not available	Input B.6 Output B.6 Output 6	Input C.6 Output C.6 Input 6	Input B.6 Output B.6 Output 6	Input C.6 Input 6
5	Input B.5 Output B.5 Output 5	Input C.5	Input B.5 Output B.5 Output 5	Input C.5	Input B.5 Output B.5 Output 5	Input C.5 Output C.5 Input 5
4	Input B.4 Output B.4 Output 4	Input C.4 Output C.4 Input 4	Input B.4 Output B.4 Output 4	Input C.4	Input B.4 Output B.4 Output 4	Input C.4 Output C.4 Input 4
3	Input B.3 Output B.3 Output 3	Input C.3 Input 3	Input B.3 Output B.3 Output 3	Output C.3	Input B.3 Output B.3 Output 3	Input C.3 Output C.3 Input 3
2	Input B.2 Output B.2 Output 2	Input C.2 Output C.2 Input 2	Input B.2 Output B.2 Output 2	Input C.2 Output C.2 Input 2	Input B.2 Output B.2 Output 2	Input C.2 Output C.2 Input 2
1	Input B.1 Output B.1 Output 1	Input C.1 Output C.1 Input 1	Input B.1 Output B.1 Output 1	Input C.1 Output C.1 Input 1	Input B.1 Output B.1 Output 1	Input C.1 Output C.1 Input 1
0	Output B.0 Output 0	Input C.0 Output C.0 Input 0	Input B.0 Output B.0 Output 0	Input C.0 Output C.0 Input 0	Input B.0 Output B.0 Output 0	Input C.0 Output C.0 Input 0

input B.1	'Make pin B.1 an input pin
output B.1	'Make pin B.1 an output pin
high B.1	'Make pin B.1 an output pin 1 and set it high
low B.1	'Make pin B.1 an output pin 1 and set it low
pinsB = %00001011	'Set port B pins 3, 1, & 0 high (must be configured as outputs)
b1 = outpinsB	'Read the state of the output port B into b1
if pinB.1 = 0 then endif	'Test the state of input pin B.1
if pinC.0 = 1 then endif	'Test the state of port C pin 0
b0 = pinsC	'Read the state of port C input pins into b0
dirsC= %00000101	'Configure port C pins 2 and 0 as output pins
pinsC = %00000011	'Set port C pins 1 and 0 high (must be configured as outputs)

TABLE 8.7 LAYOUT OF THE PINS, OUTPINS, AND DIRS VARIABLES FOR THE PICAXE-14M2

PICAXE-14M2	BIT 7	BIT 6	BIT 5	BIT 4	BIT 3	BIT 2	BIT 1	BIT 0
pinsB	-	-	pinB.5	pinB.4	pinB.3	pinB.2	pinB.1	pinB.0
pinsC	-	-	pinC.5	pinC.4	pinC.3	pinC.2	pinC.1	pinC.0
outpinsB	-	-	pinB.5	pinB.4	pinB.3	pinB.2	pinB.1	pinB.0
outpinsC	-	-	pinC.5	pinC.4	pinC.3	pinC.2	pinC.1	pinC.0
dirsB	-	-	dirB.5	dirB.4	dirB.3	dirB.2	dirB.1	dirB.0
dirsC	-	-	dirC.5	dirC.4	dirC.3	dirC.2	dirC.1	dirC.0

Interrupts

In general, the discussion of interrupts in Chapter 2 is valid for M2 series chips. The **setint** and **setintflags** commands have been enhanced to include the Boolean functions **AND** and **OR** and the updated syntax is shown here. Note that the **setintflags** command is available on X1 and X2 series chips only.

The syntax of the **setint** command is:

SETINT off

SETINT {not/and/or} condition, mask {, port}

where

off is a keyword that turns interrupts off.

not is an optional keyword on M2, X1, and X2 parts that inverts the state of the *condition*. For M2, X1, and X2 parts, **not** may be specified in conjunction with **and**.

TABLE 8.8 LAYOUT OF THE PINS, OUTPINS, AND DIRS VARIABLES FOR THE PICAXE-18M2

PICAXE-18M2	BIT 7	BIT 6	BIT 5	BIT 4	BIT 3	BIT 2	BIT 1	BIT 0
pinsB	pinB.7	pinB.6	pinB.5	pinB.4	pinB.3	pinB.2	pinB.1	pinB.0
pinsC	pinC.7	pinC.6	pinC.5	pinC.4	pinC.3	pinC.2	pinC.1	pinC.0
outpinsB	pinB.7	pinB.6	pinB.5	pinB.4	pinB.3	pinB.2	pinB.1	pinB.0
outpinsC	pinC.7	pinC.6	pinC.5	pinC.4	pinC.3	pinC.2	pinC.1	pinC.0
dirsB	dirB.7	dirB.6	dirB.5	dirB.4	dirB.3	dirB.2	dirB.1	dirB.0
dirsC	dirC.7	dirC.6	dirC.5	dirC.4	dirC.3	dirC.2	dirC.1	dirC.0

TABLE 8.9   LAYOUT OF THE PINS, OUTPINS, AND DIRS VARIABLES FOR THE PICAXE-20M2

PICAXE-20M2	BIT 7	BIT 6	BIT 5	BIT 4	BIT 3	BIT 2	BIT 1	BIT 0
pinsB	pinB.7	pinB.6	pinB.5	pinB.4	pinB.3	pinB.2	pinB.1	pinB.0
pinsC	pinC.7	pinC.6	pinC.5	pinC.4	pinC.3	pinC.2	pinC.1	pinC.0
outpinsB	pinB.7	pinB.6	pinB.5	pinB.4	pinB.3	pinB.2	pinB.1	pinB.0
outpinsC	pinC.7	pinC.6	pinC.5	pinC.4	pinC.3	pinC.2	pinC.1	pinC.0
dirsB	dirB.7	dirB.6	dirB.5	dirB.4	dirB.3	dirB.2	dirB.1	dirB.0
dirsC	dirC.7	dirC.6	dirC.5	dirC.4	dirC.3	dirC.2	dirC.1	dirC.0

and is an optional keyword that performs a logical AND on the selected inputs, **and** is the default and may be left out (i.e., all selected inputs must be true, or false if NOT is specified, for an interrupt to occur).

or is an optional keyword for M2, X1, and X2 parts that performs a logical OR on the selected inputs (i.e., one or more of the selected inputs must be true for an interrupt to occur).

condition is a variable or constant that operates in conjunction with *mask*. *Condition* specifies the state of the pins that define the interrupt. If a bit is 0 then an interrupt can occur if the corresponding pin is at logic low level and the corresponding mask bit is a 1. If a bit is 1 then an interrupt can occur if the pin is at logic high level and the corresponding mask bit is a 1.

mask is a variable or constant that defines the pins that are examined for the interrupt condition. If a mask bit is a 0 then the corresponding pin will not be examined. If a mask bit is a 1 then the corresponding pin will be examined.

port is an optional variable or constant for M2, X1, and X2 parts that defines the port that will be examined for an interrupt condition, e.g., A, B, C, D

Figure 8.3 shows how interrupt settings can be selected.

The syntax of the **setintflags** command is:

```
setintflags off
setintflags {not} flags, mask
setintflags {not/and/or} flags, mask
```

where

not is an optional keyword that inverts the state of the *condition*. For M2, X1, and X2 parts, **not** may be specified in conjunction with **and**.

and is an optional keyword that performs a logical AND on the selected inputs, **and** is the default and may be left out (i.e., all selected inputs must be true, or false if NOT is specified, for an interrupt to occur).

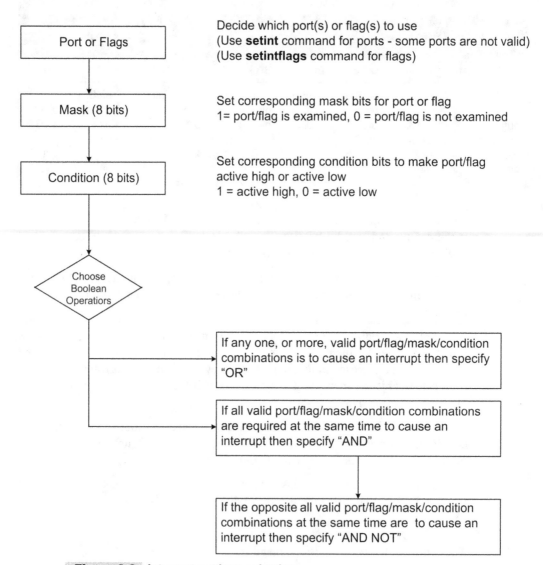

Figure 8.3 Interrupt settings selection.

or is an optional keyword for M2, X1, and X2 parts that performs a logical OR on the selected inputs (i.e., one or more of the selected inputs must be true for an interrupt to occur).

flags is a variable or constant that specifies the flags that will generate an interrupt.

mask is a variable or constant that defines the flags that are examined for the interrupt condition. If a mask bit is a 1, then the corresponding flag will be examined.

The flags are the components of the system variable "flags" and are discussed further in Chapter 2.

Figure 8.3 shows how interrupt settings can be selected.

Example:

```
setint %00010000, %00010000        'Enable interrupts on pin4 high
do
'put program code here
loop

interrupt:                          'Interrupt routine
'put interrupt routine code here
do while pin4 = 1 loop              'Optionally wait for interrupt condition to pass
pulsout 0, 1                        'Optionally send a pulse to reset the interrupt
                                      condition
setint %0010000, %0010000          'Optionally enable interrupts again
return                              'Return from interrupt routine
```

Interfacing and Input Output Techniques

In general, the interfacing and input output techniques given in Chapter 3 apply to M2 series chips for the following subjects:

Digital interfacing

Analog interfacing

Parallel and serial Interfacing

Personal computer connectivity

I2C interfacing

SPI interfacing

1-Wire interfacing

Keyboard interfacing, except that the **kbin** command is not supported by the 18M2

Infrared interfacing

Pulses

Servo motors, except that the servo pulse stream may exhibit interruptions when multitasking.

In addition, the M2 chips have the following new features:

Touch sensing

Analog output

The **readadc** and **readadc10** commands will automatically configure the port for analog input and the **adcsetup** command can be used to change the configuration of a pin from analog to digital.

Programming

The programming techniques given in Chapter 4 apply to the M2 series chips.

Compatibility with Existing M and X Series Programs

The M2 series chips maintain compatibility with existing M series and X series programs by allowing ports to be addressed either by number, or by the port-dot-bit (B.4, C.3) notation, as shown in Table 8.6.

By setting the port direction variables (dirsb, dirsc) to configure M2 Port C as input and M2 port B as output, existing M and X series programs can be run on the M2 series chips. The following line of code at the beginning of an M2 program will achieve this.

[let] dirsb = $FF   'Set all port B pins to output.

There will always be room to include this line, because M2 chips have more program memory than M and X series chips.

Subroutines

The **call** command can now be used as an alternative to the **gosub** command for all PICAXE chips. **Call** and **gosub** are functionally equivalent; subroutines are discussed in Chapter 4.

Example:

```
gosub something      'Execute subroutine "something"
call something       'Execute subroutine "something"

something:           'Subroutine "something"
   return
```

PICAXE Arithmetic and Data Conversion

The arithmetic and data conversion techniques given in Chapter 5 apply to the M2 series chips.

Parallel Task Programming (Multitasking)

The M2 series chips include the ability to run up to four tasks in parallel. Each task has equal access to the PICAXE resources, such as memory, input ports, and output ports. Executing **pause** commands in one task will not pause other task(s) that may be running, although **nap** and **sleep** commands will affect all tasks and so will any commands that use all CPU resources, such as **serin** and **readtemp**. The code for each task may be any length, and the only constraint is that the code for all tasks must fit into program memory.

The **setfreq** command is not available in programs that use multiple tasks; some background tasks, such as **servo**, may be affected. The PICAXE-18M2 byte-scratchpad is reduced in size to 128 bytes when multitasking is in use (see Fig. 8.2). This constraint does not apply to the 14M2 and 20M2.

The compiler recognizes when more than one task is specified in a program and will schedule all tasks to start at the same time. Thus, a task does not wait until the line containing the start label is executed to commence executing.

The PICAXE resources associated with parallel task processing are:

start0, **start1**, **start2**, and **start3** labels

task variable

restart, **resume**, and **suspend** commands

The **start0**, **start1**, **start2**, and **start3** labels are used to start individual tasks, the **task** variable contains the number of the task that is currently executing, and the **restart**, **resume**, and **suspend** commands are used to control the execution of tasks.

Parallel tasks may be initiated either from flowcharts in Logicator for PIC, or from program code. To initiate tasks from flowcharts in Logicator for PIC additional start boxes are placed in the flowchart. To initiate tasks from program code, the labels **start0**, **start1**, **start2**, and **start3** are placed in the program. The label **start0** is optional and if not present is assumed prior to the first line of code in a program. If present the label **start0** must be placed prior to the first line of code.

Subroutines are accessible to all tasks.

Examples of multitasking

Flowchart using Logicator for PIC is shown below:

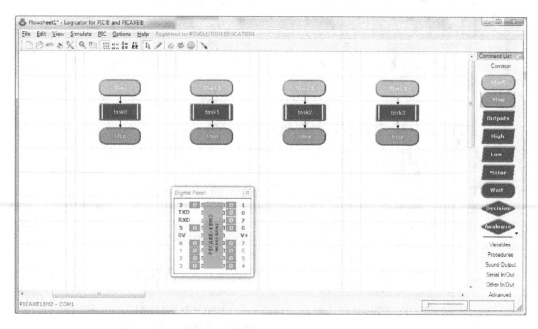

Code

start0:   'The label "start0" is optional and is assumed if not present.
                                     'If present "start0" must precede the first command
     program code for task 0
     end of task 0

start1:
     program code for task 1
     end of task 1
start2:
     program code for task 2
     end of task 2

The execution of tasks can be controlled by means of the **restart**, **resume**, and **suspend commands**. The **restart** command is used to start a task from its beginning, the **resume** command will start a suspended task from the point where it was suspended, and the **suspend** command will suspend the operation of a task. Variables are not reset by the **restart** command; thus, a task that relies on a pre-existing state for variables must perform its own initialization.

The syntax of the **restart** command is:

restart *tasknumber*

where

tasknumber is a constant or variable in the range 0–3 that specifies the task to be restarted

The **restart** command starts the task from its beginning and does not reset any variables. Thus the first command after the respective start label is the command that is executed following a **restart** command and all variables retain their previous values.

The syntax of the **resume** command is:

resume *tasknumber*

where

tasknumber is a constant or variable in the range 0–3 that specifies the task to be resumed

The **resume** command starts the task from the point where it was suspended and does not reset any variables.

The syntax of the **suspend** command is:

suspend *tasknumber*

where

tasknumber is a constant or variable in the range 0–3 that specifies the task to be suspended

If all tasks are suspended, program execution, including interrupt polling, stops and the only method of restarting the program is a hardware reset (or by reprogramming the chip).

Examples of Parallel Task Processing

The following code demonstrates multitasking by using four separate tasks to flash LED's at different rates. The code can be used with the AXE181 Touch Sensor Demonstration Board or the circuit shown later in Fig. 8.4.

```
start0:                 'Task 0 (the label 'start0' is optional and may be left out)
      do

      high B.4
      pause 512
      low B.4
      pause 512

      loop
start1:                 'Task 1
      do

      high B.5
      pause 256
      low B.5
      pause 256

      loop
```

```
start2:              'Task 2
    do

    high B.6
    pause 128
    low B.6
    pause 128

    loop

start3:              'Task 3
    do

    high B.7
    pause 64
    low B.7
    pause 64

    loop
```

Touch Sensing

Touch sensing operates by means of a change of capacitance when a finger, or other suitable object, is brought into proximity of the touch pad. The touch pad itself can be made from any conducting material, should have an area between 200 and 400 mm^2, and may be any shape. Circular and rectangular pads will be more likely to suit the profile of a finger. The touch pad must be electrically insulated from the finger and must have a single electrical connection to the PICAXE touch pin.

Other factors may influence the operation of touch pads, such as cables, especially serial programming cables, and other touch pads nearby. Each touch pin will give different readings to the others.

The commands associated with touch sensing are:

touch

touch16

The **touch** command reads a value from the specified touch-input pin and returns an 8-bit value in the specified byte variable that is proportional to the capacitance of the touch pad. The command will automatically configure the specified pin as a touch-input pin. The **adcsetup** command (variable) can be used to reconfigure the touch-input pin for digital output.

The **touch16** command reads a value from the specified touch-input pin and returns a 16-bit value in the specified word variable that is proportional to the capacitance of the touch pad. The **touch16** command is more accurate than the **touch** command and an optional configuration byte may be specified with the **touch16** command. The command will automatically configure the specified pin as a touch-input pin. The **adcsetup** command (variable) can be used to reconfigure the touch-input pin for digital output.

The syntax of the **touch** command is:

touch *channel*, *variable*

where

channel is a variable or constant specifying a touch-input pin

variable is one of the M2 byte variables

The syntax of the **touch16** command is:

touch16 [*configuration byte*], *channel*, *variable*

where

configuration byte is an optional byte constant or variable specifying a configuration value

channel is a variable or constant specifying a touch-input pin

variable is one of the M2 byte variables

The configuration values are given in the PICAXE Manual and are reproduced here for convenience.

Configuration byte

Bits 7, 6 & 5	Counter preload value 000 Oscillation count required = 256 (default) 010 Oscillation count required = 192 100 Oscillation count required = 128 110 Oscillation count required = 64 111 Oscillation count required = 32
Bits 4 & 3	00 Touch sensor oscillator is off 01 Low range (0.1 μA) – (default) 10 Medium range (1.2 μA) 11 High range (18 μA)
Bits 2, 1, & 0	Counter prescaler value 001 = divide by 4 (default)

If the configuration value is not specified, then the default value of $09 is used (oscillation count = 256, low range, and prescale divide by 4) and this is generally satisfactory for most applications.

Examples:

```
touch C.0, b0      'Read the value of touch pad C.0 into b0 as an 8-bit value.
touch16 B.3, w1    'Read the value of touch pad B.3 into w1 as a 16bit value.
```

Digital to Analog Conversion

The M2 series chips have a single digital to analog (DAC) converter that may be connected to a package pin for use by external devices. The DAC pin does not have the same output current capability as an output pin and must be buffered for most practical applications. A transistor connected in emitter-follower configuration can be used as a buffer in some applications; other applications will require more accurate results that can only be obtained by using an operational-amplifier.

The DAC is capable of being set, under program control, to 1 of 32 different levels and the maximum and minimum levels can be set to a selection of sources when the DAC is setup.

The commands associated with digital to analog conversion are:

dacsetup

daclevel

readdac

readdac10

The **dacsetup** command is used to configure the internal voltages used by the digital to analog converter. The upper and lower limits and external reference voltages can be specified.

The **daclevel** command is used to set the output level of the digital to analog converter. There are 32 levels that can be specified within the upper and lower limits specified in the most recently executed **dacsetup** command.

The **readdac** command will read the value of the DAC level as an 8-bit value. The **readdac10** command will read the value of the DAC level as a 10-bit value.

The syntax of the **dacsetup** command is:

dacsetup *config*

where

config is a constant or byte variable specifying the DAC configuration.

The configuration values are given in the PICAXE Manual and are reproduced here for convenience.

Configuration byte

bit 7	0 = DAC disabled, 1 = DAC enabled
bit 6	Not used
bit 5	0 = DAC internal only, 1 = DAC value appears on DAC output pin
bit 4	Not used
bits 3 & 2	Upper reference point
	00 = Supply voltage
	01 = External Vref+ pin – see **adcconfig** command

　　　　10 = Internal fixed-voltage reference (FVR) – see **fvrsetup** command
　　　　11 not used
　bit 1　Not used
　bit 0　Lower reference point
　　　　0 = supply 0 volt
　　　　1 = External Vref- pin – see **adcconfig** command

The syntax of the **daclevel** command is:

　daclevel *level*

where

　level is a constant or variable specifying the DAC output level in the range 0–31

　There are 32 possible values for *level* and they specify 32 levels of output for the DAC, where 0 is the lower reference point specified in the **dacsetup** command, and 31 is the upper reference point specified in the **dacsetup** command. Using the fixed-voltage reference as the upper reference point allows the DAC to output precision voltages, although the DAC output pin will usually need to be buffered when connected to an external device.

The syntax of the **readdac** command is:

　readdac *variable*

where

　variable is any of the PICAXE variables.

The syntax of the **readdac10** command is:

　readdac10 *wordvariable*

where

　wordvariable is any of the PICAXE word variables.

　Examples:

```
dacsetup %10100000        'Configure DAC on with external output and
                          'upper level = supply voltage and
                          'lower level = ground.
daclevel 0                'Set the DAC level to 0 (minimum value)
daclevel15                'Set the DAC level to 15
daclevel 31               'Set the DAC level to 31 (maximum value)
```

Fixed-Voltage Reference

The M2 chips have an internal fixed-voltage reference (FVR) that can be set to different voltages or turned off under program control.

The command **fvrsetup** will set the value of the internal fixed-voltage reference. The syntax of the **fvrsetup** command is:

fvrsetup *value*

where

value is one of:

OFF	Turns the internal fixed-voltage reference off
FVR1024	1.024 V
FVR2048	2.048 V
FVR4096	4.096 V – requires a 4.5 or 5 V supply

Example:

fvrsetup FVR2048   Sets the internal fixed-voltage reference to 2.048 V.

Elapsed Seconds Timer

The PICAXE M2 series chips have a 16-bit elapsed time counter that is incremented once per second. The timer is set to zero at power on or when a reset command is executed and may be controlled by program commands.

The commands associated with the elapsed seconds timer are:

disabletime

enabletime

The **disabletime** command disables the elapsed seconds timer, and the **enabletime** command enables the elapsed seconds timer.

The syntax of the **disabletime** and **enabletime** commands are:

disabletime

enabletime

The value of the elapsed seconds timer can be read from, or written to, the system variable **time** (s_w7).

Examples:

w0 = time	'Read the elapsed seconds timer
time = 0	'Reload the elapsed seconds timer
disabletime	'Stop the elapsed seconds timer from incrementing
enabletime	'Allow the elapsed seconds timer to increment every second

SR Latch

The SR latch is a hardware flip-flop that can be set and reset by commands or events. It is available on the PICAXE M2 series and PICAXE 20X2 chips. The SR latch output (SRQ), SR latch inverted output (SRNQ – 20X2 only), and SR latch input (SRI) can be configured to be available at a package pin for use by external devices. The SR latch input (SRI) can be configured to set or reset the SR latch by means of the **srlatch** command.

The commands associated with the SR latch are:

srlatch

srset

srreset

The **srlatch** command is used to configure the SR latch, the **srset** command will set the SR latch (SRQ high), and the **srreset** command will reset the SR latch (SRQ low).

The syntax of the **srlatch** command is:

srlatch *config1, config2*

where

config1 is a variable or constant that specifies configuration settings

config2 is a variable or constant that specifies configuration settings

The values for *config1* and *config2* are given in the PICAXE Manual and are reproduced here for convenience.

Configuration byte 1 (*config1*)

bit 7	1 = SR latch active, 0 SR latch inactive
bits 6, 5, & 4	SR Clock Divider bits – sets latch clock frequency

	000 divide by 4	(0.25μs @ 16 MHz, 0.5 μs @ 8 MHz, 1 μs @ 4 MHz)
	001 divide by 8	(0.5 μs @ 16 MHz, 1 μs @ 8 MHz, 2 μs @ 4 MHz)
	010 divide by 16	(1 μs @ 16 MHz, 2 μs @ 8 MHz, 4 μs @ 4 MHz)
	011 divide by 32	(2 μs @ 16 MHz, 4 μs @ 8 MHz, 8 μs @ 4 MHz)
	100 divide by 64	(4 μs @ 16 MHz, 8 μs @ 8 MHz, 16 μs @ 4 MHz)
	101 divide by 128	(8 μs @ 16 MHz, 16 μs @ 8 MHz, 32 μs @ 4 MHz)
	110 divide by 256	(16 μs @ 16 MHz, 32 μs @ 8 MHz, 64 μs @ 4 MHz)
	111 divide by 512	(32 μs @ 16 MHz, 64 μs @ 8 MHz, 128 μs @ 4 MHz)

bit 3	1 = SR latch output available at SRQ pin (must be configured as output)
	0 = SR latch not available at SRQ pin
bit 2	1 = SR latch inverted output available at SRNQ pin (must be output pin)
	0 = SR latch inverted output not available at SRNQ pin
bit 1	Not used
bit 0	Not used

Configuration byte 2 (*config2*)

For PICAXE-20X2

bit 7	1 = SR latch is set by HINT1, see **hintsetup** command
bit 6	1 = SR latch is set by internal clock, see config1 bits 6, 5, & 4
bit 5	1 = SR latch is set by C2 comparator, see **compsetup** command
bit 4	1 = SR latch is set by C1 comparator, see **compsetup** command
bit 3	1 = SR latch is reset by HINT1, see **hintsetup** command
bit 2	1 = SR latch is reset by internal clock, see config1 bits 6,5 & 4
bit 1	1 = SR latch is reset by C2 comparator, see **compsetup** command
bit 0	1 = SR latch is reset by C1 comparator, see **compsetup** command

"Set" and "reset" bits should not be set at the same time. If a set and reset are issued at the same time, "reset" takes precedence over "set."

For PICAXE M2 Series

bit 7	1 = SR latch is set by a logic high applied to SRI pin
bit 6	1 = SR latch is set by internal clock, see config1 bits 6, 5, & 4
bit 5	Not used
bit 4	Not used
bit 3	1 = SR latch is reset by a logic high applied to SRI pin
bit 2	1 = SR latch is reset by internal clock, see config1 bits 6, 5, & 4
bit 1	Not used
bit 0	Not used

"Set" and "reset" bits should not be set at the same time. If a set and reset are issued at the same time, "reset" takes precedence over "set."

The syntax of the **srset** command is:

srset

The syntax of the **srreset** command is:

srreset

An example can be found in the SR latch demonstration experiment later in this chapter.

M2 Experiments

The AXE181 Touch Sensor Demo Board is used for most of the M2 experiments and the circuit is shown in Fig. 8.4. You can make up this circuit yourself, using the guidelines for touch pad construction (see earlier in this chapter), and achieve the same results.

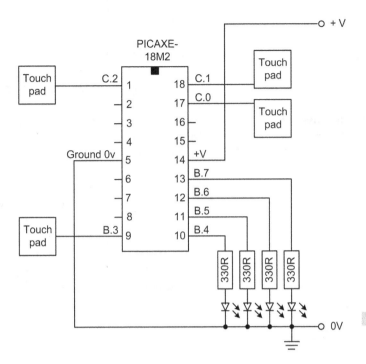

Figure 8.4 Circuit for the touch sensor experiment.

Touch Sensing

This experiment makes use of the AXE181 Touch Sensor Demo Board to demonstrate touch sensing. In this experiment four touch pads and four LEDs are connected to a PICAXE-18M2. When the touch pads are touched the corresponding LED lights. The code for the touch sensor demonstration is:

```
'PICAXE- 18M2 Touch Sensor Demonstration
#picaxe 18m2
```

symbol touchthresh = 1000	'Touch difference threshold, values between 1000 and 2500 work well
symbol touch0p = w0	'Touch C.0 previous value
symbol touch1p = w1	'Touch C.1 previous value
symbol touch2p = w2	'Touch C.2 previous value
symbol touch3p = w3	'Touch B.3 previous value
symbol touchval = w4	'Current value for current touch sensor
symbol touchdiff = w5	'Current sensor difference reading

```
         'Initialize previous touch values
         touch16 C.0, touch0p
         touch16 C.1, touch1p
         touch16 C.2, touch2p
         touch16 B.3, touch3p
```

```
do                              'Main loop
'Sensor C.0
touch16 C.0, touchval           'Read touch pad value

if touchval > touch0p then      'Compare the current reading with the previous
                                   reading
    touchdiff = touchval - touch0p    'Calculate difference for an increase in value
    if touchdiff > touchthresh then   'If the difference is above the threshold value
        high B.4                'turn the LED on
        endif
    else
    touchdiff = touch0p - touchval    'Calculate difference for a decrease in value
if touchdiff > touchthresh then   'If the difference is above the threshold value
        low B.4                 'Turn the LED off
        endif
endif
touch0p = touchval              'Save the current value as previous


'Sensor C.1
touch16 C.1, touchval           'Read touch pad value

if touchval > touch1p then      'Compare the current reading with the
                                   previous reading
    touchdiff = touchval - touch1p    'Calculate difference for an increase in value
    if touchdiff > touchthresh then   'If the difference is above the threshold value
        high B.5                'turn the LED on
        endif
    else
    touchdiff = touch1p - touchval    'Calculate difference for a decrease in value
if touchdiff > touchthresh then   'If the difference is above the threshold value
        low B.5                 'Turn the LED off
        endif
endif
touch1p = touchval              'Save the current value as previous

'Sensor C.2
touch16 C.2, touchval           'Read touch pad value

if touchval > touch2p then      'Compare the current reading with the previous
                                   reading
    touchdiff = touchval - touch2p    'Calculate difference for an increase in value
    if touchdiff > touchthresh then   'If the difference is above the threshold value
        high B.6                'turn the LED on
        endif
    else
    touchdiff = touch2p - touchval    'Calculate difference for a decrease in value
```

```
if touchdiff > touchthresh then        'If the difference is above the threshold value
    low B.6                            'Turn the LED off
    endif
endif
touch2p = touchval                     'Save the current value as previous


'Sensor B.3
touch16 B.3, touchval                  'Read touch pad value

if touchval > touch3p then             'Compare the current reading with the previous
                                       reading
    touchdiff = touchval - touch3p     'Calculate difference for an increase in value
    if touchdiff > touchthresh then    'If the difference is above the threshold value
    high B.7                           'turn the LED on
        endif
    else
    touchdiff = touch3p - touchval     'Calculate difference for a decrease in value
if touchdiff > touchthresh then        'If the difference is above the threshold value
    low B.7                            'Turn the LED off
    endif
endif
touch3p = touchval                     'Save the current value as previous


loop
```

CODE DESCRIPTION

The code compares the current value of the touch sensor reading with the previous value of the touch sensor reading. If there is a significant increase in the reading, the corresponding LED is turned on; if there is a significant decrease in the reading, the corresponding LED is turned off. This method tends to compensate for the differences in readings between different touch ports and any variations in the readings that might occur over time.

Digital to Analog Conversion

In this experiment the digital to analog converter in the PICAXE-18M2 and the fixed-voltage reference are used to produce an audio tone that has a peak-to-peak level of 2.048 V. A 10-K resistor in conjunction with a 0.22-μF capacitor forms a simple filter that helps to remove unwanted signal components. The signal is then fed to the input of an audio amplifier connected to a speaker. Almost any audio amplifier can be used and an LM386 is used for the audio amplifier in this experiment.

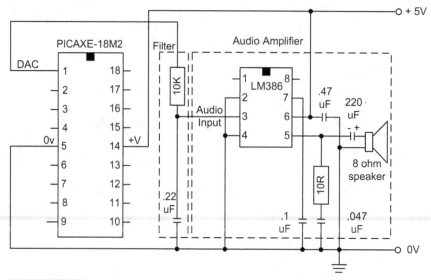

Figure 8.5 Circuit for the digital to analog conversion experiment.

The circuit is shown in Fig. 8.5.

The code for the digital to analog experiment is:

```
'Digital to analog conversion producing audio tone

#picaxe 18m2

        setfreq m32             'Run at maximum clock speed to produce highest
                                    frequency tone
                                'Lower clock frequencies produce lower frequency tones

        fvrsetup FVR2048        'Configure fixed-voltage reference for 2.048 V
        dacsetup %10101000      'Configure DAC on with external output, and
                                'upper level = fixed-voltage reference, and
                                'lower level = ground.

        do                      'Main loop

        daclevel 0              'Use a series of values that approximate a sine wave.
        daclevel 5
        daclevel 16
        daclevel 27
        daclevel 31
        daclevel 27
        daclevel 16
        daclevel 5

        loop                    'Repeat indefinitely
```

CODE DESCRIPTION

The code first sets the PICAXE clock frequency to 32 MHz to produce the highest frequency audio tone; lower clock frequencies will produce lower frequency tones. The fixed-voltage reference is configured to 2.048 V and the DAC is then configured for DAC on, external output on, upper level = fixed-voltage reference, and lower level = supply ground.

The code then enters an endless loop that sets the DAC levels to values that approximate a sine wave with a peak-to-peak voltage equal to the fixed-voltage reference. The audio volume can be changed by varying the values to produce different output levels, or by setting the fixed-voltage reference to a different voltage.

Minute Timer

This experiment uses the AXE181 Touch Sensor Demo Board or the circuit in Fig. 8.4 to demonstrate the use of the elapsed seconds timer to flash a LED every minute.

The code for the minute timer is:

```
'Minute timer

'Flash a LED every 60 s

#picaxe 18m2

    do

    if time >= 60 then          'If the elapsed time counter is 60 or more
        time = 0                'Reset the elapsed time counter
        pulsout B.4, 50000      'Turn a LED on for half a second
    endif

    loop                        'Repeat indefinitely
```

To produce a beep every minute, substitute a piezo speaker for the LED and change the code to:

```
'Minute timer

'Beep every 60 s

#picaxe 18m2

    do

    if time >= 60 then          'If the elapsed time counter is 60 or more
        time = 0                'Reset the elapsed time counter
        sound B.4, (100,50)     'Beep for half a second
    endif

    loop                        'Repeat indefinitely
```

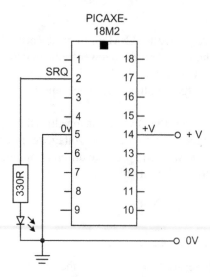

Figure 8.6 Circuit for the SR latch demonstration program.

SR Latch Demonstration

A LED is connected to the SRQ output of a PICAXE-18M2 and code is used to flash the LED. The circuit is shown in Fig. 8.6.

The code to demonstrate the SR latch for PICAXE-18M2 is:

```
'SR Latch test
#picaxe 18m2

srlatch %10001000, %00000000    'SR active, SR latch output available at SRQ pin
pinsC = pinsC or %00001000      'Make C.3 (SRQ) an output pin

do

srset                           'Set the SR latch, SRQ = high
pause 500
srreset                         'Reset the SR latch, SRQ = low
pause 500

loop
```

A

CIRCUIT SYMBOLS

The subject of electronics could fill an entire volume by itself and, in many cases, has done so. In the next few pages I have tried to summarize some of the principles that are needed to gain a better understanding of how the circuits in this book work.

Voltage, Current, and Resistance

Voltage is the unit of measure of *electromotive force* (EMF), which is defined as the amount of energy required to move a unit of electrical charge between two points. The

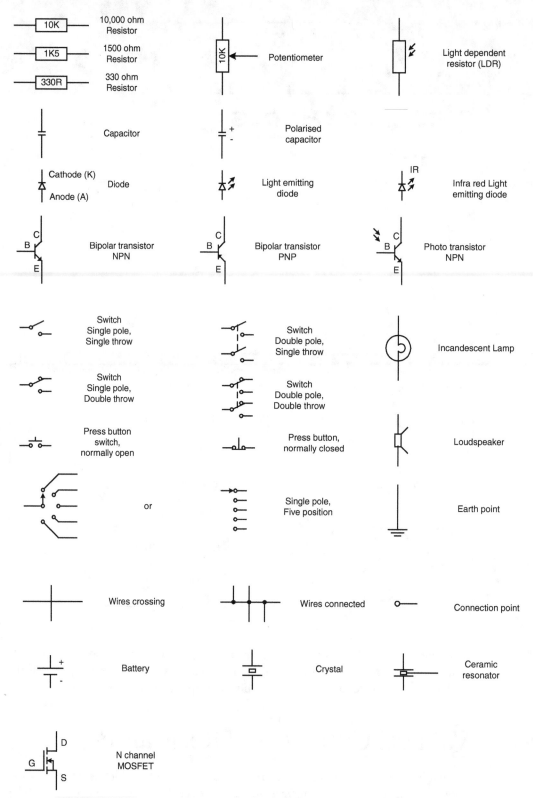

Figure A.1 Circuit symbols used in the book.

TABLE A.1   ASCII CHARACTER SET

LOW NIBBLE	0	1	2	3	4	5	6	7	8	9	A	B	C	D	E	F
High nibble																
0	NUL	SOH	STX	ETX	EOT	ENQ	ACK	BEL	BS	HT	LF	VT	FF	CR	SO	SI
1	DLE	DC1	DC2	DC3	DC4	NAK	SYN	ETB	CAN	EM	SUB	ESC	FS	GS	RS	US
2	Space	!	"	#	$	%	&	'	(	)	*	+	,	-	.	/
3	0	1	2	3	4	5	6	7	8	9	:	;	<	=	>	?
4	@	A	B	C	D	E	F	G	H	I	J	K	L	M	N	O
5	P	Q	R	S	T	U	V	W	X	Y	Z	[	\	]	^	_
6	'	a	b	c	d	e	f	g	h	i	j	k	l	m	n	o
7	p	q	r	s	t	u	v	w	x	y	z	{			}	~	DEL

unit of EMF is the volt, often abbreviated V. In electronics, the volt can be a large unit of EMF and, in practice, EMF is often measured in millivolts (10^{-3} V), abbreviated as mV.

Current is the amount of electrical charge that passes a particular point in a given time. The unit of current is the ampere, often abbreviated amp or A. In electronics, the ampere is a large unit of current and, in practice, current is often measured in milliamps (10^{-3} amps), abbreviated as mA. Milliamps may be too large a unit in some applications and the term microamps (10^{-6} amps), abbreviated as μA, may be used.

Resistance is the ability of a material to impede the flow of electric current. The unit of resistance is the ohm, often abbreviated as Ω. In electronics, the ohm is a small unit of resistance and, in practice, resistance is often measured in kilo-ohms (10^{3} Ω), abbreviated as Kohm or K, or megohms (10^{6} Ω), abbreviated as Mohm or meg.

Voltage can be regarded as electrical pressure, and current as the amount of electricity that is flowing. To explain it simply, an analogy with water is often used. Suppose that we have a water tank that is, for example, 4 m high (about 12 ft). If we start at the surface and move down, the water pressure will increase with depth. The water pressure is analogous to electrical pressure or voltage. If we now put a hole in the bottom of the water tank and connect a small pipe, for example, 6 mm (about 1/4 in) in diameter, some water will flow through the pipe. If we now put another hole in the bottom of the tank, this time 300 mm (about 12 in) in diameter, then a much larger volume of water will flow.

The volume of water flowing is analogous to electrical current. The larger pipe allows more water to flow because it offers less impediment to the water than the smaller pipe. The impediment to water flow is analogous to electrical resistance.

Ohm's law defines the relationship between voltage, current, and resistance in an electrical circuit. Ohm's law states that the current flowing in a circuit is directly proportional to the electrical pressure and inversely proportional to the resistance and is expressed by the following formula:

$$I = E/R$$

where

I is the current in Amperes (A)

E is the applied EMF in Volts (V)

R is the resistance in Ohms (Ω)

By transposition,

$$E = IR \text{ and } R = E/I$$

Voltage sources may be *direct current* (DC) or *alternating current* (AC). Direct current flows in a single direction in a circuit and may be a constant voltage or a fluctuating voltage. Alternating current periodically reverses direction and the voltage varies between zero and positive and negative peak values. The voltage of an alternating current source is often given as the *root mean square* (rms) value rather than the peak value, because this is equivalent to the same direct current voltage in power equations. For a sine wave, the peak AC voltage is equal to the rms value multiplied by the square root of 2 (approximately 1.41).

Batteries are a common source of direct current and the power mains is a common source of alternating current. Direct current can also be obtained from alternating current by using a combination of rectifier, filter, and voltage regulator.

Voltage sources may be connected in series to give a higher voltage by connecting the positive terminal of one source to the negative terminal of the next (in phase for AC). The total voltage of series-connected power sources is equal to the sum of the individual voltages. Batteries are often connected in series to give a higher voltage. For example, a 12-V car battery consists of six individual cells producing around 2 V each.

Voltage sources may also be connected in parallel, although some care should be taken if doing so. When connecting voltage sources in parallel, all sources should be the same voltage with all the positive terminals connected together and all the negative

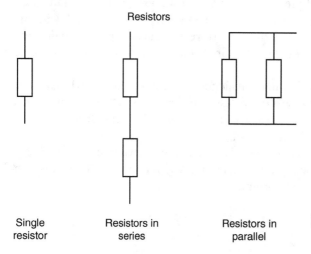

Single resistor   Resistors in series   Resistors in parallel

Figure A.2 **Resistors in series and parallel.**

terminals connected together (in phase for AC). All the parallel connected sources should have the same internal resistance (or impedance for AC).

Power

Power in an electrical circuit results in heat, which appears as a temperature rise in the component that dissipates the power. This heat must be removed, usually by radiation into the air, to prevent components from overheating and "burning out." Some components generate very little heat and thus do not show a noticeable increase in temperature; other components may generate larger amounts of heat and care must be taken to ensure that the component can dissipate all the heat generated within it. Power is measured in watts (W) and is given by the formula:

$$P = EI$$

where

P is power in Watts (W)

E is applied EMF in Volts (V)

I is current flowing in Amperes (A)

Power can also be calculated from voltage or current and resistance. From Ohm's law we know that $E = IR$ and $I = E/R$; these expressions can be substituted in the power equation to give the following:

$$P = E^2/R \text{ and } P = I^2 R$$

where

P is power in Watts (W)

E is applied EMF in Volts (V)

I is the current flowing in Amperes (A)

R is the resistance in Ohms (Ω)

Resistors

Resistors are two-terminal devices that are not polarity sensitive. They are rated in terms of their resistance, power dissipation capability, and material of construction. Resistors are commonly made from carbon film, metal film, or resistance wire, and have a maximum voltage rating, although this is rarely an issue for low-voltage circuits.

The resistance value is usually marked on the resistor body by means of colored bands, although some resistors may have the resistance value printed on the body of the resistor. The resistance of a resistor is not always exactly the same as the value that is marked on it. The difference that can occur between the marked value and the actual value is called the *tolerance*. Resistors are manufactured with tolerances of 5, 1, and 0.1 percent, and other tolerances are possible. If a resistor has a tolerance of 5 percent, then the actual resistance will be within ± 5 percent of the value marked on the resistor. The same principle applies to other tolerances. In digital circuits the tolerance of a resistor is usually not critical and 5% is generally adequate. However, in analog circuits, tolerances are more important and 0.1% may not be sufficient for some applications.

Resistors must be able to dissipate all the heat generated within them or they can suffer damage and may burn out. Resistors are commonly manufactured with ratings of 0.25, 0.5, 1.0, and 5 W, although other values are possible. Higher-power resistors are usually physically larger than low-power resistors in order to provide more surface area to dissipate heat. The amount of power dissipated in a resistor can be calculated from the following formula:

$$P = EI$$

where

P is power in Watts (W)

E is the applied EMF in Volts (V)

I is the current flowing in Amperes (A)

The power rating of a resistor should be greater than the calculated power in order to keep the surface temperature low. Running a resistor at, or close to, its maximum power rating can result in high surface temperature that may cause premature failure, often evidenced by the resistance value changing with time.

Resistors are not the only components that generate heat. Integrated circuits (including PICAXE chips) generate heat within them and it is important to ensure that their power dissipation ratings are not exceeded. For this reason it is a good idea to place resistors in series with LEDs that are connected to PICAXE output ports, because most of the power will be dissipated in the resistor rather than the chip. By contrast, if the internal current limiting of the PICAXE chip is used to limit the output current, the power will be dissipated within the chip.

RESISTORS IN SERIES AND PARALLEL

When two or more resistors are connected in series, the total resistance is equal to the sum of the individual resistors. When two or more resistors are connected in parallel, the total resistance is less than the resistance of the lowest resistance value (see Fig. A.2).

For parallel connected resistors, the total resistance can be calculated from the formula:

$$1/R_T = 1/R_1 + 1/R_2 + 1/R_n$$

where

R_T is the total resistance

R_1 is resistance value 1

R_2 is resistance value 2

R_n is resistance value n

For two resistors, this formula can be simplified to

$$R_T = R_1 * R_2/(R_1 + R_2)$$

If two or more resistors are connected in series, the voltage drop across each resistor is directly proportional to the resistance value and this principle can be used to construct a voltage divider. The voltage drop across any resistor in a voltage divider can be calculated from the following formula:

$$V_R = R/R_{TOT} * V_{TOT}$$

where

V_R is the voltage drop across an individual resistor

R is the resistance of an individual resistor

R_{TOT} is the total resistance

V_{TOT} is the total voltage applied to the voltage divider (see Fig. A.3)

For example, if resistors of 100, 200, and 300 Ω are connected in series across a 6 V supply, a 1 V drop will take place across the 100 Ω resistor, a 2 V drop across the 200 Ω resistor, and a 3 V drop across the 300 Ω resistor.

On many occasions it is necessary to introduce resistors into a circuit in order to limit the amount of current that can flow. Limiting the current flowing through a *light-emitting diode* (LED) and limiting the current flowing through the base-emitter circuit of a bipolar transistor are two examples.

Examples

What value resistor is needed to limit the current to 20 mA for a red LED that is connected to a 5 V source?

The value of the current-limiting resistor can be calculated from Ohm's law. The red LED drops 1.2 V, leaving 3.8 V to be dropped by the resistor. Thus, the resistance is given by $3.8/(20 * 10^{-3}) = 190$ Ω. The next highest preferred value in the 5 percent range is 220 Ω and this will produce slightly less than 20 mA, which won't matter much for most practical applications.

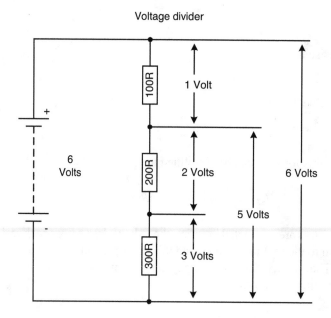

Voltage divider

Figure A.3 Voltage divider.

The power dissipated in the current-limiting resistor should now be calculated so that a suitably rated resistor can be used. From Ohm's law and the power formula, we know that power is equal to E^2/R. For a 220 Ω resistor dropping 3.8 V, the power is 0.066 W, which is well within the capabilities of a 0.25 W resistor. Thus, a 220 Ω, 0.25 W carbon film resistor would be suitable for this application.

What value resistor is needed to limit the base current of a bipolar transistor to 1 mA when it is connected to a *transistor-transistor logic* (TTL) output?

The value of the current-limiting resistor can be calculated from Ohm's law. A TTL output delivers 5 V (if the supply voltage is 5 V) and the base emitter junction of a bipolar transistor drops about 0.5 V, leaving 4.5 V to be dropped by the resistor. Thus, the resistance is given by $4.5/(1 * 10^{-3}) = 4500 \Omega$. The next highest preferred value in the 5 percent range is 4700 Ω and this will produce slightly less than 1 mA, which should not be a problem for most applications.

The power dissipated in the current-limiting resistor should now be calculated so that a suitably rated resistor can be used. From Ohm's law and the power formula, we know that power is equal to I^2R. For a 4700 Ω resistor passing 1 mA, the power is 0.0047 W, which is well within the capabilities of a 0.25-W resistor. Thus, a 4700 Ω, 0.25-W carbon film resistor would be suitable for this application.

READING THE VALUE OF A COLOR-CODED RESISTOR

Color-coded resistors typically have four or five color bands, depending on their tolerance (some specialized resistors may have more or less bands) as illustrated in Fig. A.4.

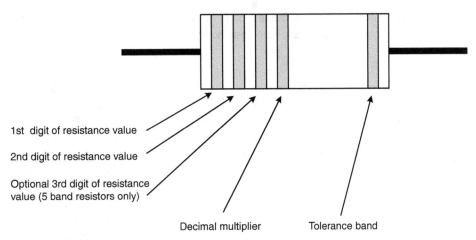

1st digit of resistance value

2nd digit of resistance value

Optional 3rd digit of resistance
value (5 band resistors only)

Decimal multiplier   Tolerance band

Figure A.4 Resistor color bands.

To read the value of a color-coded resistor, follow these steps:

1. Locate the first digit (see Fig. A.4).
2. The second band is the second digit.
3. If the resistor has five color bands, then the third band is the third digit, and the fourth band is the decimal multiplier.
4. If the resistor has four color bands, then the third band is the decimal multiplier.
5. The resistance value is found by:
   a. The digit corresponding to the color of the first band followed by
   b. The digit corresponding to the color of the second band followed by
   c. The digit corresponding to the color of the third band (if there is one)
   d. Multiplied by the multiplier corresponding to the color of the third or fourth band

TABLE A.2   RESISTOR COLOR CODE

COLOR	NUMERIC VALUE	MULTIPLIER VALUE	TOLERANCE
Silver		10^{-2}	10%
Gold		10^{-1}	5%
Black	0	10^{0}	
Brown	1	10^{1}	1%
Red	2	10^{2}	2%
Orange	3	10^{3}	
Yellow	4	10^{4}	
Green	5	10^{5}	
Blue	6	10^{6}	
Violet (Purple)	7	10^{7}	
Grey	8	10^{8}	
White	9	10^{9}	

The tolerance band is the last band on the resistor. Silver and gold are used only in the decimal multiplier and tolerance bands.

Examples

Brown, red, gold, gold	$1.2\ \Omega$, 5% $(1, 2, 10^{-1}, 5\%)$
Brown, black, black, gold	$10\ \Omega$, 5% $(1, 0, 10^{0}, 5\%)$
Orange, orange, brown, gold	$330\ \Omega$, 5% $(3, 3, 10^{1}, 5\%)$
Green, blue, brown, gold	$560\ \Omega$, 5% $(5, 6, 10^{1}, 5\%)$
Brown, black, red, gold	$1,000\ \Omega$ (1K), 5% $(1, 0, 10^{2}, 5\%)$
Yellow, violet, red, gold	$4,700\ \Omega$ (4.7K), 5% $(4, 7, 10^{2}, 5\%)$
Brown, black, orange, gold	$10,000\ \Omega$ (10K), 5% $(1, 0, 10^{3}, 5\%)$
Orange, orange, black, black, brown	$330\ \Omega$, 1% $(3, 3, 0, 10^{0}, 1\%)$
Brown, black, black, brown, brown	$1,000\ \Omega$, 1% $(1, 0, 0, 10^{1}, 1\%)$
Brown, black, black, red, brown	$10,000\ \Omega$, 1% $(1, 0, 0, 10^{2}, 1\%)$

Capacitors

Capacitors are two-terminal devices that are made from two conducting materials separated from each other by a nonconducting material. Capacitors may be polarized or nonpolarized. Polarized capacitors, often called electrolytic capacitors, must have a DC bias, of the correct polarity, connected to their terminals in order to properly function.

Capacitors are rated in terms of their capacitance and the maximum voltage that may be applied to them. The internal resistance and inductance may also be important in some applications. The unit of capacitance is the farad, which is defined as 1 coulomb/volt.

A farad is a very large value of capacitance, and in practice capacitors are usually measured in microfarads (10^{-6} farad), nanofarads (10^{-9} farad), or picofarads (10^{-12} farad).

When two or more capacitors are connected in series, the total capacitance is less than the capacitance of the smallest capacitor and can be calculated from the following formula:

$$1/C_{TOT} = 1/C_1 + 1/C_2 + 1/C_n$$

where

C_{TOT} is the total capacitance

C_1 is capacitance value 1

C_2 is capacitance value 2

C_n is capacitance value n

For two capacitors, this formula can be simplified to

$$C_{TOT} = C_1 * C_2/(C_1 + C_2)$$

Capacitors

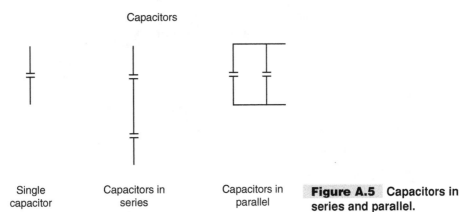

| Single capacitor | Capacitors in series | Capacitors in parallel | **Figure A.5** Capacitors in series and parallel. |

When two or more capacitors are connected in parallel, the total capacitance is equal to the sum of the values of all the capacitors (see Fig. A.5).

Capacitors allow alternating current to pass through while preventing direct current from flowing. Reactance is the apparent resistance of a capacitor to alternating current, which varies with frequency. Capacitive reactance is higher at low frequencies and is given by the formula:

$$X_C = 1/(2\pi f C)$$

where

X_C is the capacitive reactance in ohms

π is the constant pi (approximately 3.141)

f is the frequency in hertz

C is the capacitance in farads

When reactance and resistance are combined in a circuit, the apparent resistance is known as *impedance*.

Capacitors can be used in conjunction with a resistor to create a time delay, because it takes a period of time for a capacitor to become charged or discharged (see Fig. A.6).

RC network

Figure A.6 Resistor–capacitor network.

The time for a capacitor connected in series with a resistor to charge or discharge to 63 percent of the applied voltage is given by the formula:

$$T = CR$$

where

T is the time delay in seconds

C is the capacitance in farads

R is the resistance in ohms

Inductors

Inductors are constructed from a conductive material, such as copper wire, which is wound in a coil or spiral (some specialized inductors may consist of less than a single turn). To increase the inductance, the coil may have a core of iron or ferrite. The unit of inductance is the henry and is defined as 1 weber/ampere. Inductors may be connected in series or parallel, although it is unusual to do so in most applications.

Inductors allow the passage of direct current while resisting the passage of alternating current. Reactance is the apparent resistance to alternating current, which varies with frequency. Inductive reactance is lower at low frequencies and is calculated by the formula:

$$X_L = 2\pi fL$$

where

X_L is the inductive reactance in ohms

π is the constant pi (approximately 3.141)

f is the frequency in hertz

L is the inductance in henries

When a current passes through an inductor, a magnetic field builds up and the inductor is said to be charged. If the current is removed, the magnetic field will rapidly collapse, creating a voltage across the inductor terminals known as *back EMF*. The back EMF has the opposite polarity to the charging voltage and is usually large enough to damage semiconductor devices. When an inductive device such as a relay coil, motor, or solenoid is being switched by a transistor, it is necessary to take precaution to prevent the back EMF from destroying the transistor. It is common practice to place a reverse-biased diode across the inductor to absorb the back EMF (see Fig. A.7).

Back EMF Protection

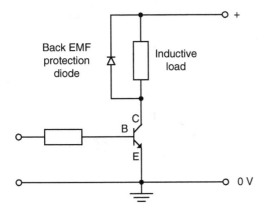

Figure A.7 **Back EMF protection diode.**

Semiconductors

Semiconductors are materials that are neither electrical conductors nor electrical insulators. By definition, semiconductors are materials that have conductivity between 10^{-7} and 10^3 mho/cm. Germanium and silicon are two of the better known semiconducting materials; silicon is the material that is most often used for manufacturing diodes, transistors, and *integrated circuits* (ICs).

Silicon by itself is not a particularly useful substance, but when it is changed into P- and N-type silicon, by means of a process known as doping, it becomes extremely useful. The doping process adds controlled quantities of other substances, such as boron and phosphorous, to the silicon to change its conduction characteristics.

A diode is a device that allows electric current to pass in one direction only. It is made by placing pieces of N- and P-type semiconductor material, usually silicon (sometimes germanium or gallium arsenide for LEDs), next to each other. A diode has two terminals known as the anode and cathode, and the point where the N- and P-type material meets is known as the junction. Diodes are usually rated in terms of the maximum reverse voltage they can sustain without breaking down and the maximum current they can pass.

When a diode is conducting, it is said to be forward-biased and when it is not conducting it is said to be reversed-biased. A forward-biased diode will have a small voltage drop between the anode and cathode, typically between 0.5 and 1.2 V, depending on the amount of current flowing (see Fig. A.8).

| K | Cathode |
| A | Anode |

Diode

Figure A.8 **Diode.**

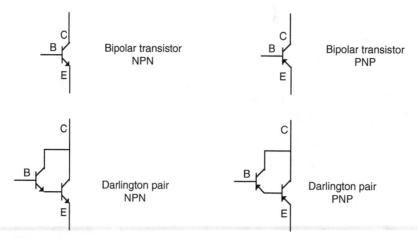

Figure A.9 **Bipolar transistor symbols**

Bipolar transistors are devices that can amplify current. They are made by placing a piece of N-type silicon between two pieces of P-type silicon (*positive-negative-positive* [PNP] transistor) or a piece of P-type silicon between two pieces of N-type silicon (*negative-positive-negative* [NPN] transistor). Bipolar transistors have three terminals, which are known as the base, emitter, and collector.

A small current flowing in the base emitter circuit of a bipolar transistor causes a larger current to flow in the collector emitter circuit, thus providing current amplification. The base emitter junction of a bipolar transistor forms a diode, which drops around 0.5 to 0.7 V, depending on the current flowing.

Bipolar transistors are usually rated in terms of the maximum voltage they can sustain between the collector and the emitter without breaking down, the maximum current that can flow in the collector emitter circuit, the current gain (beta or Hfe), and the maximum amount of power they can dissipate.

Some transistors are made from a combination of two transistors and these are known as a Darlington pair or Darlington transistors. Darlington transistors behave in the same way as single transistors except that the current gain is equal to the product of the current gains of the individual transistors used. The base emitter voltage drop is 1.0 to 1.4 V, depending on the current flowing (see Fig. A.9).

Darlington transistors may be manufactured with the two transistors in the same package or may be constructed from two individual transistors.

Optical Devices

Optical devices include the following:

- Incandescent lamps
- Light-emitting diodes (LEDs)
- Light-dependent resistors (LDRs)
- Photo diodes
- Photo transistors

Incandescent lamps are made from an electrically conductive filament, usually tung-sten wire. An electric current is passed through the filament for the specific purpose of heating the wire until it is white hot thus, producing light. The filament must be prevented from coming into contact with oxygen, which would cause it to burn. It is enclosed in a glass bulb that is evacuated or contains an inert gas at low pressure. There is currently a worldwide trend to phase out incandescent lamps in favor of light-emitting diodes and fluorescent lamps.

LEDs are made from a semiconductor material, such as gallium arsenide, which gives off light when current is flowing through it. LEDs are two-terminal devices that allow current to pass in one direction only.

LDRs are made from cadmium sulfide, which changes its resistance when exposed to light. Cadmium sulfide photoresistors are two-terminal devices acting as resistors. They do not oppose current flow in a particular direction as do diodes.

Certain diodes and transistors can change their conduction characteristics if their junctions are exposed to light. They are known as photo diodes and photo transistors. Photo diodes have two terminals, and photo transistors may have two or three terminals, because some photo transistors do not have a base terminal.

Relays

Relays are electrically operated switches that use an electromagnet to operate switch contacts. Relays may be monostable or bistable. Bistable relays have two electromagnets (coils); one is the set coil and the other is the reset coil. Bistable relays maintain their state when power is removed and monostable relays always revert to their at-rest, or normal, state when power is removed (see Fig. A.10).

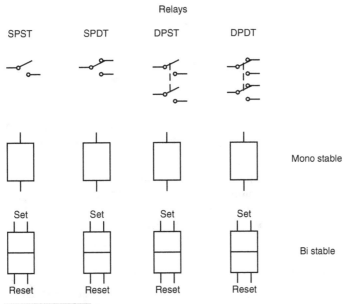

Figure A.10 **Relays.**

Hard Reset

Circumstances may occur when the Programming Editor is unable to communicate with a PICAXE while downloading a program. In these circumstances, a program download can usually be achieved by performing a hard reset.

The hard reset procedure requires that the PICAXE be placed in a reset state when program download commences either by being powered off or by holding the reset line low. When program download has commenced, the reset condition must be removed before a download timeout occurs.

Troubleshooting

If your circuit does not work for any apparent reason, try the following:

- Connecting the PICAXE to a regulated 5 V supply using short leads.
- Connecting the PICAXE to a 4.5 V supply made up of three alkaline cells.
- Connecting a 0.1-μF or 1-μF monolithic bypass capacitor to the PICAXE supply pins as close as possible to the PICAXE chip.
- Connecting a 0.1-μF monolithic bypass capacitor to the supply pins of any other chips that are being used in the circuit.
- Running relays, motors, lamps, speakers, or solenoids from a separate power supply.
- Disconnecting the programming cable after programming.
- Leaving the programming circuit (10- and 22-K resistors) connected permanently.
- Connecting the frames of electric motors, stepper motors, solenoids, or other devices to ground.

Substitute Components

Substitutes may be used for the following components:

- **1N4001/4**: The 1N4001/4 is a general-purpose power diode rated at 1 A. Almost any power diode with a voltage rating of 50 V or more and a current rating of 1 A or more can be substituted.
- **1N5819**: This is a general-purpose Schottky power diode. Almost any Schottky power diode with a voltage rating of 30 V or more can be substituted.
- **BC548**: The BC548 is a general-purpose, low-power NPN transistor, and almost any general-purpose NPN transistor can be substituted. The important ratings are VCE 30 V or more, IC 100 mA or more, Hfe 80 or more, and power dissipation 500 mW or more.
- **BD139**: The BD139 is a general-purpose NPN power transistor, and almost any general-purpose NPN power transistor can be substituted. The important

ratings are V_{CE} 80 V or more, I_C 1.5 A or more, Hfe 40 or more, and P_{TOT} 8 W or more.

■ **BD681**: The BD681 is a general-purpose NPN Darlington power transistor, and almost any general-purpose NPN Darlington transistor can be substituted. The important ratings are V_{CE} 80 V or more, I_C 4 A or more, Hfe 750 or more, and P_{TOT} 20 W or more.

■ **TSOP4838**: The TSOP4838 is an infrared receiver module designed for remote control systems operating at 38 kHz. Other infrared receiver modules designed to operate at 38 kHz can often be substituted, although it may be necessary to vary the value of the resistor between pin 3 and the positive supply for successful operation.

■ **7404**: The 7404 is a TTL Hex inverter and any TTL-compatible inverter can be substituted, including 74LS04 and 74HC04. The 74xx00 can also be substituted if both inputs of each gate are connected together.

■ **L293D**: The L293D is a quadruple half-H driver with internal diodes. The SN754410 is a pin-compatible alternative to the L293

Data

PICAXE-08M pin connections

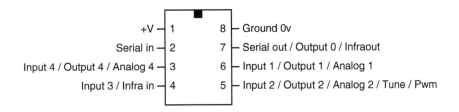

PICAXE-14M pin connections

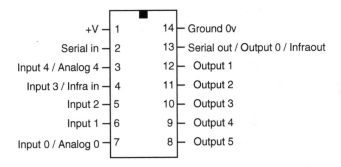

PICAXE-18M pin connections

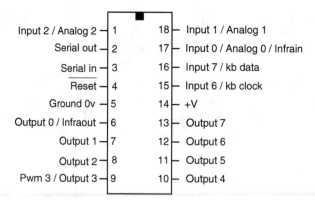

PICAXE-18X pin connections

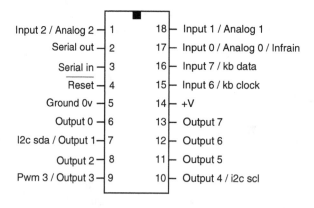

PICAXE-20M pin connections

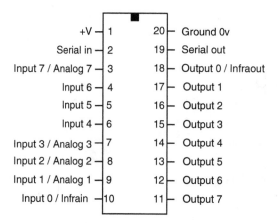

PICAXE-28X1 pin connections

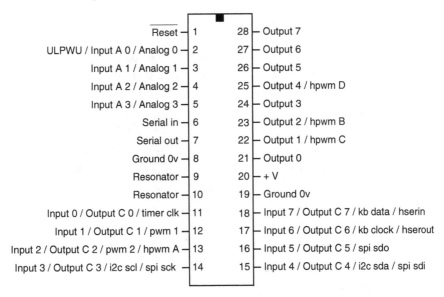

```
                    Reset ─┤ 1      28 ├─ Output 7
  ULPWU / Input A 0 / Analog 0 ─┤ 2      27 ├─ Output 6
         Input A 1 / Analog 1 ─┤ 3      26 ├─ Output 5
         Input A 2 / Analog 2 ─┤ 4      25 ├─ Output 4 / hpwm D
         Input A 3 / Analog 3 ─┤ 5      24 ├─ Output 3
                   Serial in ─┤ 6      23 ├─ Output 2 / hpwm B
                  Serial out ─┤ 7      22 ├─ Output 1 / hpwm C
                  Ground 0v ─┤ 8      21 ├─ Output 0
                  Resonator ─┤ 9      20 ├─ + V
                  Resonator ─┤ 10     19 ├─ Ground 0v
   Input 0 / Output C 0 / timer clk ─┤ 11     18 ├─ Input 7 / Output C 7 / kb data / hserin
   Input 1 / Output C 1 / pwm 1 ─┤ 12     17 ├─ Input 6 / Output C 6 / kb clock / hserout
Input 2 / Output C 2 / pwm 2 / hpwm A ─┤ 13  16 ├─ Input 5 / Output C 5 / spi sdo
  Input 3 / Output C 3 / i2c scl / spi sck ─┤ 14  15 ├─ Input 4 / Output C 4 / i2c sda / spi sdi
```

PICAXE-40X1 pin connections

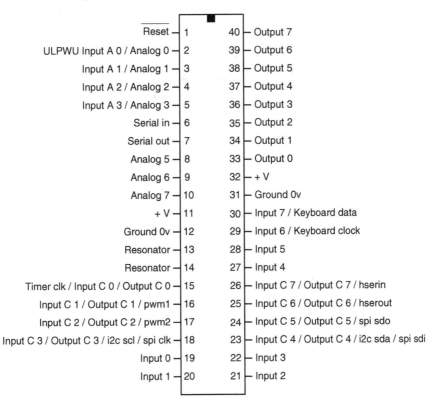

```
                    Reset ─┤ 1      40 ├─ Output 7
    ULPWU Input A 0 / Analog 0 ─┤ 2      39 ├─ Output 6
         Input A 1 / Analog 1 ─┤ 3      38 ├─ Output 5
         Input A 2 / Analog 2 ─┤ 4      37 ├─ Output 4
         Input A 3 / Analog 3 ─┤ 5      36 ├─ Output 3
                   Serial in ─┤ 6      35 ├─ Output 2
                  Serial out ─┤ 7      34 ├─ Output 1
                   Analog 5 ─┤ 8      33 ├─ Output 0
                   Analog 6 ─┤ 9      32 ├─ + V
                   Analog 7 ─┤ 10     31 ├─ Ground 0v
                       + V ─┤ 11     30 ├─ Input 7 / Keyboard data
                  Ground 0v ─┤ 12     29 ├─ Input 6 / Keyboard clock
                  Resonator ─┤ 13     28 ├─ Input 5
                  Resonator ─┤ 14     27 ├─ Input 4
   Timer clk / Input C 0 / Output C 0 ─┤ 15  26 ├─ Input C 7 / Output C 7 / hserin
   Input C 1 / Output C 1 / pwm1 ─┤ 16  25 ├─ Input C 6 / Output C 6 / hserout
   Input C 2 / Output C 2 / pwm2 ─┤ 17  24 ├─ Input C 5 / Output C 5 / spi sdo
Input C 3 / Output C 3 / i2c scl / spi clk ─┤ 18  23 ├─ Input C 4 / Output C 4 / i2c sda / spi sdi
                     Input 0 ─┤ 19     22 ├─ Input 3
                     Input 1 ─┤ 20     21 ├─ Input 2
```

PICAXE-28X1 pin connections

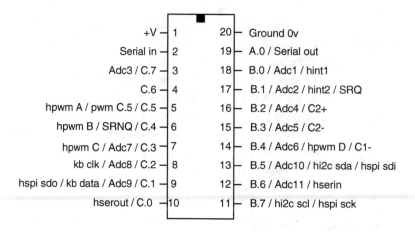

+V	—	1	20	—	Ground 0v
Serial in	—	2	19	—	A.0 / Serial out
Adc3 / C.7	—	3	18	—	B.0 / Adc1 / hint1
C.6	—	4	17	—	B.1 / Adc2 / hint2 / SRQ
hpwm A / pwm C.5 / C.5	—	5	16	—	B.2 / Adc4 / C2+
hpwm B / SRNQ / C.4	—	6	15	—	B.3 / Adc5 / C2-
hpwm C / Adc7 / C.3	—	7	14	—	B.4 / Adc6 / hpwm D / C1-
kb clk / Adc8 / C.2	—	8	13	—	B.5 / Adc10 / hi2c sda / hspi sdi
hspi sdo / kb data / Adc9 / C.1	—	9	12	—	B.6 / Adc11 / hserin
hserout / C.0	—	10	11	—	B.7 / hi2c scl / hspi sck

PICAXE-28X2 pin connections

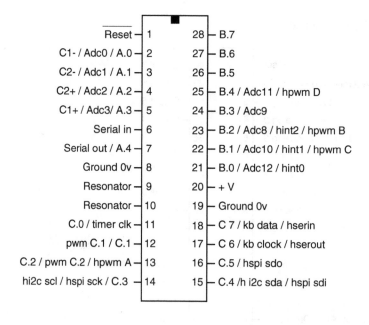

Reset	—	1	28	—	B.7
C1- / Adc0 / A.0	—	2	27	—	B.6
C2- / Adc1 / A.1	—	3	26	—	B.5
C2+ / Adc2 / A.2	—	4	25	—	B.4 / Adc11 / hpwm D
C1+ / Adc3/ A.3	—	5	24	—	B.3 / Adc9
Serial in	—	6	23	—	B.2 / Adc8 / hint2 / hpwm B
Serial out / A.4	—	7	22	—	B.1 / Adc10 / hint1 / hpwm C
Ground 0v	—	8	21	—	B.0 / Adc12 / hint0
Resonator	—	9	20	—	+ V
Resonator	—	10	19	—	Ground 0v
C.0 / timer clk	—	11	18	—	C 7 / kb data / hserin
pwm C.1 / C.1	—	12	17	—	C 6 / kb clock / hserout
C.2 / pwm C.2 / hpwm A	—	13	16	—	C.5 / hspi sdo
hi2c scl / hspi sck / C.3	—	14	15	—	C.4 /h i2c sda / hspi sdi

PICAXE-40X2 pin connections

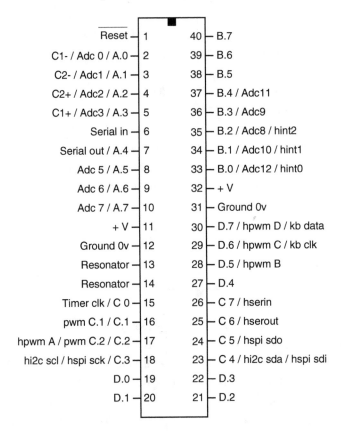

Reset — 1	40 — B.7	
C1- / Adc 0 / A.0 — 2	39 — B.6	
C2- / Adc1 / A.1 — 3	38 — B.5	
C2+ / Adc2 / A.2 — 4	37 — B.4 / Adc11	
C1+ / Adc3 / A.3 — 5	36 — B.3 / Adc9	
Serial in — 6	35 — B.2 / Adc8 / hint2	
Serial out / A.4 — 7	34 — B.1 / Adc10 / hint1	
Adc 5 / A.5 — 8	33 — B.0 / Adc12 / hint0	
Adc 6 / A.6 — 9	32 — + V	
Adc 7 / A.7 — 10	31 — Ground 0v	
+ V — 11	30 — D.7 / hpwm D / kb data	
Ground 0v — 12	29 — D.6 / hpwm C / kb clk	
Resonator — 13	28 — D.5 / hpwm B	
Resonator — 14	27 — D.4	
Timer clk / C 0 — 15	26 — C 7 / hserin	
pwm C.1 / C.1 — 16	25 — C 6 / hserout	
hpwm A / pwm C.2 / C.2 — 17	24 — C 5 / hspi sdo	
hi2c scl / hspi sck / C.3 — 18	23 — C 4 / hi2c sda / hspi sdi	
D.0 — 19	22 — D.3	
D.1 — 20	21 — D.2	

PICAXE-14M2 pin connections

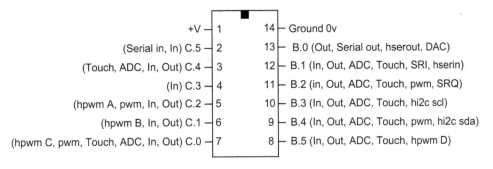

+V — 1	14 — Ground 0v
(Serial in, In) C.5 — 2	13 — B.0 (Out, Serial out, hserout, DAC)
(Touch, ADC, In, Out) C.4 — 3	12 — B.1 (In, Out, ADC, Touch, SRI, hserin)
(In) C.3 — 4	11 — B.2 (in, Out, ADC, Touch, pwm, SRQ)
(hpwm A, pwm, In, Out) C.2 — 5	10 — B.3 (In, Out, ADC, Touch, hi2c scl)
(hpwm B, In, Out) C.1 — 6	9 — B.4 (In, Out, ADC, Touch, pwm, hi2c sda)
(hpwm C, pwm, Touch, ADC, In, Out) C.0 — 7	8 — B.5 (In, Out, ADC, Touch, hpwm D)

PICAXE-18M2 pin connections

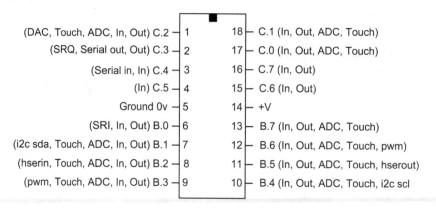

(DAC, Touch, ADC, In, Out) C.2	1	18	C.1 (In, Out, ADC, Touch)
(SRQ, Serial out, Out) C.3	2	17	C.0 (In, Out, ADC, Touch)
(Serial in, In) C.4	3	16	C.7 (In, Out)
(In) C.5	4	15	C.6 (In, Out)
Ground 0v	5	14	+V
(SRI, In, Out) B.0	6	13	B.7 (In, Out, ADC, Touch)
(i2c sda, Touch, ADC, In, Out) B.1	7	12	B.6 (In, Out, ADC, Touch, pwm)
(hserin, Touch, ADC, In, Out) B.2	8	11	B.5 (In, Out, ADC, Touch, hserout)
(pwm, Touch, ADC, In, Out) B.3	9	10	B.4 (In, Out, ADC, Touch, i2c scl

PICAXE-20M2 pin connections

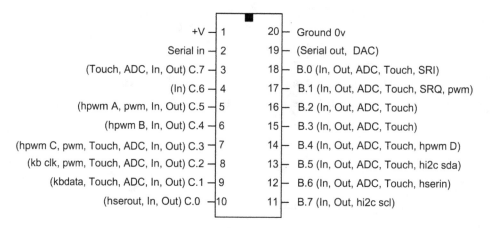

+V	1	20	Ground 0v
Serial in	2	19	(Serial out, DAC)
(Touch, ADC, In, Out) C.7	3	18	B.0 (In, Out, ADC, Touch, SRI)
(In) C.6	4	17	B.1 (In, Out, ADC, Touch, SRQ, pwm)
(hpwm A, pwm, In, Out) C.5	5	16	B.2 (In, Out, ADC, Touch)
(hpwm B, In, Out) C.4	6	15	B.3 (In, Out, ADC, Touch)
(hpwm C, pwm, Touch, ADC, In, Out) C.3	7	14	B.4 (In, Out, ADC, Touch, hpwm D)
(kb clk, pwm, Touch, ADC, In, Out) C.2	8	13	B.5 (In, Out, ADC, Touch, hi2c sda)
(kbdata, Touch, ADC, In, Out) C.1	9	12	B.6 (In, Out, ADC, Touch, hserin)
(hserout, In, Out) C.0	10	11	B.7 (In, Out, hi2c scl)

INDEX

CPSIA information can be obtained
at www.ICGtesting.com
Printed in the USA
LVHW101359290920
667409LV00012B/548

9 780071 745543